TYPICAL FLIES

SERIES III

TYPICAL FLIES

A PHOTOGRAPHIC ATLAS OF DIPTERA

BY

E. K. PEARCE

F.L.S., F.E.S.

SERIES III

CAMBRIDGE

AT THE UNIVERSITY PRESS

1928

CAMBRIDGE
UNIVERSITY PRESS

University Printing House, Cambridge CB2 8BS, United Kingdom

Cambridge University Press is part of the University of Cambridge.

It furthers the University's mission by disseminating knowledge in the pursuit of education, learning and research at the highest international levels of excellence.

www.cambridge.org
Information on this title: www.cambridge.org/9781107461604

First published 1928
First paperback edition 2014

A catalogue record for this publication is available from the British Library

ISBN 978-1-107-46160-4 Paperback

To

THE MEMORY

OF

MY MOTHER & FATHER

PREFACE TO THE THIRD SERIES

IN the preparation of this third Series I have again to thank Professor Theobald for many of the notes elucidating the life-histories. I am also indebted to Major E. E. Austen, D.S.O., F.R.S. and to Mr Edwards, of the British Museum for the determination of several specimens and for some important notes; to Dr Hugh Scott for permission to obtain a photograph of the *Atherix Ibis* cluster in the Museum at Cambridge; and to Mr B. Harwood of Sudbury, Mr Waddington of Bournemouth, the late Mr Carter of Monifieth, Dundee, Mr Grimshaw of Edinburgh, Dr Haines of Dorset, Mr N. D. F. Pearce of Grantchester and Mr H. Jones of the New Forest for the provision of specimens.

I must have photographed some 500 specimens in the course of many years. Only those who have handled these wonderful insects can understand the great difficulties of doing them full justice; and incidental blemishes will, it is hoped, be overlooked.

The sequence of species has followed that given in the late Mr Verrall's *List of British Diptera* and Brauer's Classification has been given by Professor Theobald's permission; all suggestions as to method of collection are indicated in the Prefaces to the first and second series.

E. K. PEARCE

MORDEN, WAREHAM, DORSET
May 14, 1928

PREFACE TO THE FIRST SERIES

THE study of Diptera (two-winged flies) is rendered peculiarly difficult by the lack of elementary treatises on the subject. Certain groups are fully treated in the two large (and costly) volumes published by the late Mr Verrall, there are sundry scattered papers in various magazines, and monographs (such as Lowne's on the Blowfly); but there is nothing to compare with the numerous manuals dealing with Lepidoptera and Coleoptera, to name two orders only. This little book does not claim to fill the gap, but it is hoped that it may be of some use to the beginner, and attract attention to an order which possesses great interest, and is moreover of much economic importance.

It is chiefly a picture book, as pictures appeal more to the eye than many pages of letterpress; and an important dipterous character—the venation of the wings—can be rendered with fidelity in a photograph.

I have found it difficult to obtain specimens set sufficiently flat for photographic reproduction; since, in photographing on the enlarged scale required, no amount of "stopping down" will produce an image sharp all over, unless the subject be fairly in one plane: in addition to this, some species when set and dried shrivel up, and give but a poor idea of their appearance when fresh. This of course chiefly applies to the *bodies* of flies, the wings and legs are not so affected.

Flies may be taken with the usual entomological net, preferably a green one, as less likely to cause alarm than a white one. Mosquito netting, which may be dyed the required colour, is much better than green leno. The net should be fairly large, but light and easily managed, as many flies are very swift and strong on the wing. When caught, the fly may be transferred to a glass bottomed entomological box: a good supply should be carried, and it is better that only one specimen be placed in a box. On returning home the flies may be killed in a laurel bottle, care being taken that the leaves do not become mildewed, which would probably ruin the specimens; a circular piece of white blotting paper should be placed over the leaves, and frequently renewed. Flies should remain in the bottle till they are thoroughly relaxed, which will require a day or two; if left too long they become rotten and easily break whilst setting. Narrow boards, such as are used for the smallest lepidoptera, will be suitable for large and medium sized flies; small ones may be set on strips of flat cork, covered with thin white paper. Entomological forceps will be needed to insert the pin in the thorax of the fly; I prefer these curved, as they are also useful for moving pinned specimens. If the flies are to be photographed the pin must be cut off as short as possible above the thorax, and the cut end blackened with a touch of "matt black." No. 20 pins will be useful for most flies, though the large species require something stronger, whilst the very small ones, if pinned at all, require the finest silver pins obtainable. Taylor, New Hall Works, Birmingham, will supply a sample card of pins. For setting, strips and triangles of stiff writing paper, to hold legs and wings in place, and a stiff sable paint-brush, a few handled bristles and a fine needle or two, also handled, will do all that is required: a lens is indispensable in setting small flies, and may with advantage be mounted on a simple stand to leave both hands free. Some flies, especially the

Tachinidae, are very brittle: care must be taken in manipulating them. Culicidae should be set *and photographed* as quickly as possible, they very soon shrivel. Other flies may remain about ten days on the setting boards. As to numbers, half a dozen specimens should be ample, both sexes being represented, where possible. Fewer will often have to suffice with rare species, and for purposes of photography *one* well set specimen would be sufficient, were it not for the ever present risk of damage in moving from the store box: the slightest touch or jar will often cause the loss of a leg or antenna, and the attempt to replace these is seldom successful.

Store boxes may be had in many sizes (10 x 8 inches is as good as any); whatever size is used should be adhered to, as far as possible, for the sake of uniformity. They should be carefully examined for mites, a great enemy to the collector; even new boxes are sometimes contaminated. In sending flies by post it is well to use two boxes, pinning them well into the inner, and supporting them by extra pins if possible; then packing the box with shavings inside a larger one. The label will of course be *tied* on. When finally pinning flies into the store box it is essential to use a small label giving date and locality, which can be pinned, written side down, by the same pin as the specimen. The name, etc. of the fly is written on a second label and pinned behind it in the box: the sex should be marked, where known, and a number added to correspond with that in a notebook, where fuller details may be recorded. Flies should be stored in a cool dry place, free from accidental jars and careless handling. Naphthalin wrapped in a piece of net should be pinned in a corner of the box as a guard against mites, the great enemy of the dipterist as of the entomologist in general.

The chief season for collecting in this country is from March till October, the sunny forenoon being the best time. Windy days are very unproductive. Even well-known and good localities are sometimes a blank, from causes we do not know, for flies seem very capricious in their habits. But, like other creatures, they have special haunts where they may usually be found at the proper season, and where they may be expected to occur if carefully searched for. Considerable experience in their habits and localities is needed by the collector. Generally speaking, umbelliferous plants, also bramble, hawthorn and ivy bloom seem to be most attractive. Flies often settle upon gate-posts, railings, and tree-trunks, especially if wounded or decayed. Others frequent salt-marshes and swamps, ponds and river-sides; whilst heath-lands, sheepruns, bare hot sandy areas and commons attract others. Horse and cattle droppings and decomposing animal and vegetable matter are well-known baits for many species. Others attack living animals, not excepting man; and certain flies prey on insects and spiders. Should horses or cattle be approached for the purpose of taking flies, much care must be exercised, as a net will generally stampede them; it is difficult to employ it to advantage under such conditions.

This little book has received the kind encouragement of many entomologists, among whom I may mention Professor Nuttall and Mr Warburton, both of Cambridge. Much practical help in the selection of species, and information as to types selected and their larvae, has been afforded by the kindness of Professor Theobald, whose assistance, it is hoped, has added greatly to the utility of this book. Thanks are also due to Mr Harwood, of Sudbury, Suffolk, for some of the fine specimens of diptera which he has furnished for the photographs herewith presented. It has not always been possible to do them justice, owing to the difficulties previously noted as besetting the photographer. Mr H. Waddington kindly supplied some fine microscopic slides. The author's brother, Mr N. D. F. Pearce, has also helped with the illustrations, as to the success of which the reader must be left to judge. Acknowledgement has been made in every case, it is thought, where help has been received: and if this effort is successful it is hoped that it may be some day supplemented by a further series of pictures, to fill a few gaps that were unavoidable in the present volume. The life history (ovum, larva, pupa) of many of the species shown is yet to be traced by entomologists. Measurements are given in every case in millimetres (25 mm. = 1 inch), the first dimension being the length of the fly, and the second the expanse of wings. For various reasons it has not been found possible to reproduce the specimens on one uniform scale: the same difficulty was met with by Dr Michael in illustrating the Oribatidae.

E. K. PEARCE

BOURNEMOUTH
June 1915

BRAUER'S CLASSIFICATION OF DIPTERA

* An asterisk intimates that the family will be found in Series I.
† An obelus that it is illustrated in Series II.
§ A section-mark that it occurs in Series III.

Sub-order 1. **ORTHORRHAPHA**

Larva with a distinct head. Pupa obtected.

The adult escapes from the pupal skin by a straight dorsal slit which may be transverse but is more usually longitudinal. Imago lacks the frontal lunule and ptilinum.

Sub-order 2. **CYCLORRHAPHA**

Larva without any distinct head. The Pupa coarctate.

The adult escapes from the puparium through a more or less round opening at the anterior end. Frontal lunule present; ptilinum usually present.

Sub-order 1. **ORTHORRHAPHA**

Section I. NEMATOCERA

Antennae long and thread-like, composed of many similar or very similar segments. The maxillary palpi usually elongate and flexible of from 2 to 5 segments. Second long vein often forked.

Section II. BRACHYCERA

Antennae usually of three segments, the third usually elongated and sometimes composed of a number of indistinct sub-segments and often bearing a style or arista. Maxillary palpi of 1 to 2 segments, not flexible. Second long vein not forked. Squamae completely concealing the halteres.

1. **THE ORTHORRHAPHA.** Section I. NEMATOCERA contain the following families:

*1. [1] Pulicidae (Fleas).
*2. Cecidomyidae (Gall Midges).
§†*3. Mycetophilidae (Fungus Gnats).
†*4. Bibionidae (Fever Flies, St Mark's Flies).
†*5. Simuliidae (Sand Flies).
§*6. Chironomidae (Midges).
7. Orphnephilidae.
†8. Psychodidae (Owl Midges).
§†*9. Culicidae (Mosquitoes).
§†10. Dixidae.
§*11. Ptychopteridae (False Daddy Long Legs).
§*12. Limnobiidae (False Daddy Long Legs).
§†*13. Tipulidae (True Daddy Long Legs).
§*14. Rhyphidae (Window Flies).

Section II. BRACHYCERA

§†*15. Stratiomyidae (Chameleon Flies).
§†*16. Tabanidae (Gad Flies).
§*17. Leptidae (Leptis Flies).
§†*18. Asilidae (Robber Flies).
§†*19. Bombylidae.
§*20. Therevidae.
†21. Scenopinidae.
§†22. Cyrtidae.
§†*23. Empidae (Empis Flies).
§†*24. Dolichopodidae.
†25. Lonchopteridae.

2. **THE CYCLORRHAPHA.** Section I. ASCHIZA

Frontal lunule more or less indefinite; no frontal suture.

§*26. Platypezidae.
†27. Pipunculidae.
§†*28. Syrphidae (Hover Flies).

[1] These are by some raised to the rank of an order called *Aphaniptera* or *Siphonaptera*, but there is no reason for separating the Fleas or Pulicidae from the Diptera.

Section II. SCHIZOPHORA

Frontal lunule and frontal suture marked.

Sub-section A. Muscoidea

Produce ova as a rule.

Sub-section B. Pupipara

Produce fully matured larvae.

Sub-section A. Muscoidea
Series *a*. *Acalyptrata*

Squamae small, not concealing the halteres.

Series *b*. *Calyptrata*

Squamae concealing the halteres.

Section II. SCHIZOPHORA

Sub-section A. Muscoidea. Series *a*. *Acalyptrata*

§*29. Conopidae.
§†*30. Cordyluridae.
§31. Phycodromidae.
§32. Helomyzidae.
§33. Heteroneuridae.
§†*34. Sciomyzidae.
§†35. Psilidae.
§36. Micropezidae.
§†37. Ortalidae.
§†*38. Trypetidae.
§†39. Lonchaeidae.
§40. Sapromyzidae.
†41. Opomyzidae.
§42. Sepsidae.
§†43. Piophilidae (Cheese Flies).
44. Geomyzidae.
§45. Ephydridae.
†46. Drosophilidae.
§*47. Chloropidae (Gout Flies)
48. Milichidae.
49. Agromyzidae.

†50. Phytomyzidae.
51. Astiadae.
†52. Borboridae.
†53. Phoridae.

Sub-section A. MUSCOIDEA. Series *b*. *Calyptrata*

†*54. Oestridae (Warble Flies).
§†*55. [1] Tachinidae (Tachina Flies).
§†*56. Muscidae (House Flies, etc.).
§†*57. Anthomyidae (Root-feeding Maggots, etc.).

Sub-section B. PUPIPARA

§†*58. Hippoboscidae (Forest Flies).
†59. Braulidae (Bee Flies).
†60. Nycteribiidae (Bat Flies).

[1] The Sarcophaginae and Dexinae are sometimes separated from the Tachinidae as two separate families.

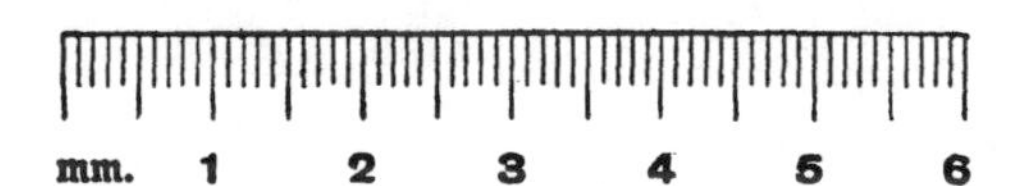

BIBLIOGRAPHY

The following books may be of use to Students and should be consulted in important libraries if out of print.

G. H. Verrall. *List of British Diptera.* 2nd Edition. 2*s.*

—— *British Flies.* Masterly volumes on (1) Stratiomyidæ, (2) Syrphidæ. 30*s.* vol.

E. E. Austen. *Illustrations of British Blood-Sucking Flies.* 1906. 25*s.*

Lang. *Handbook of British Mosquitoes.* British Museum. 20*s.*

Wingate. "List of Durham Diptera." *Trans. Nat. Hist. Soc. of Northumberland.* Vol. II (new series), 1906. (Out of print.) Sometimes can still be acquired. Very useful.

Grimshaw. *A Guide to the Literature of British Diptera.* 1917. 2*s.* (Of the Author) Royal Scottish Museum, Edinburgh.

British Museum. "Instructions for Collectors," and "Map of Distribution of Anopheles": also "The House Fly as a danger to Health," "Mosquitoes."

Poulton. "Predacious Insects and their Prey." *Trans. Ent. Soc. London*, Part V, 1906.

Theobald. *Monograph of the Culicidæ, or Mosquitoes.* 5 vols. 1901–1910.

—— *An Account of British Flies* (short series).

—— *Reports on Economic Zoology.* Insect Pests.

F. W. Edwards. "Notes on British *Mycetophilidae.*" *Trans. Ent. Soc. London*, Part II, 1913.

W. Lundbeck. *Diptera Danica* (in English).

Miall. *The Natural History of Aquatic Insects.* (Macmillan.)

Williston. *North American Diptera.* The 1896 edition and the illustrated 1908 edition out of print.

Dr H. Scott. "Notes on Nycteribiidae." *Parasitology*, Vol. IX.

Bagnall and Harrison. "A Preliminary Catalogue of British Cecidomyidae." *Trans. Ent. Soc. London*, Part IV, 1917.

Osten-Sacken. *Comparative Chaetotaxy*: on the arrangement of bristles of Diptera: see also Williston's *North American Diptera.*

F. W. Edwards. "British Limnobiidae." *Trans. Ent. Soc. London*, Part II, 1921.

—— "A Revision of the Mosquitoes." *Bull. Ent. Res.* Vol. XII, 1921.

Many others (such as G. H. Verrall, Eaton, J. E. Collins) may be learnt in Grimshaw's *Guide* already named—a full and concise epitome of Diptera literature. Many of these Transactions must be sought in Public or Museum Libraries, or those of the Learned Societies in London and the provinces.

Meigen. *Zweiflügeligen Insekten*, 1818–1838.

Macquart. *Diptères*, 1834 (ancient classics of Dipterology) often alluded to, can be seen as above.

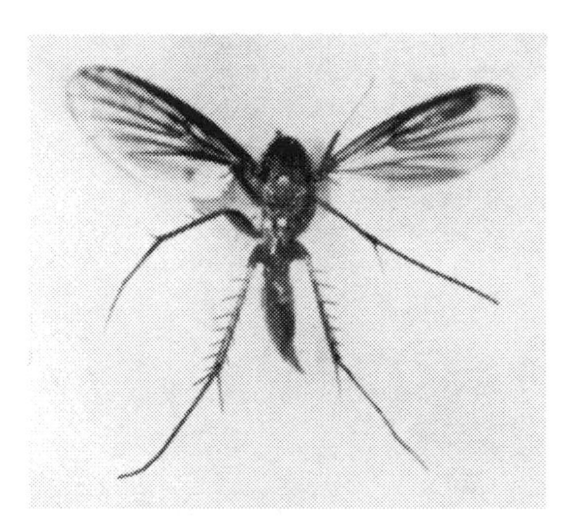

Fig. 1. *Mycetophila fungorum* De Geer. ♀(?) 5×9 mm. (*punctata* Mg.) (Determined F. W. Edwards.) A luteous brown fly, brownish wings. New Forest, Adams. The larvae of this gnat feed on fungi, especially *Boletus luteus*.

Fig. 2. *Mycetophila fungorum* De Geer. (Closed wings.) ♂(?) 5 mm.

(*a*)

(*b*)

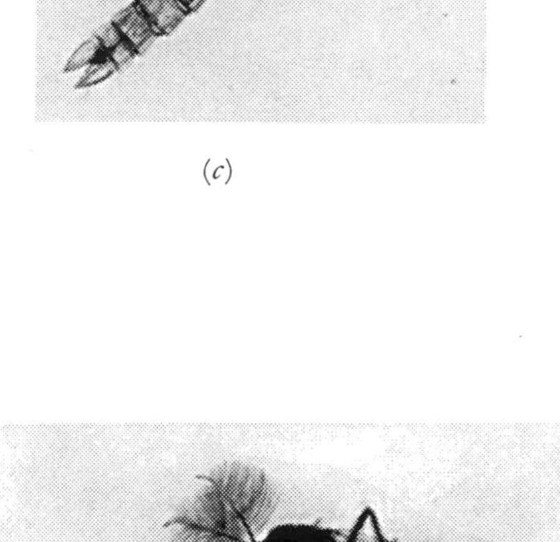

(*c*)

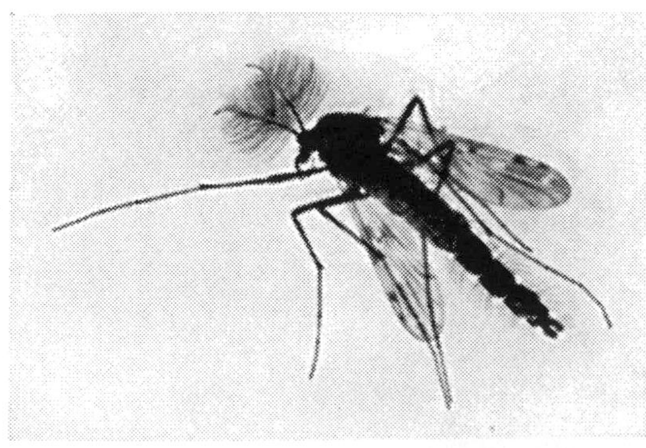

(*d*) ♂

(*d*) ♀

Fig. 3. *Tanypus* Mg. (*a*) Larva, $5\frac{1}{2}$ mm.; (*b*) pupa emerging from larva case. Entire length $6\frac{1}{2}$ mm. (pupa only 4 mm.); (*c*) imago emerging from pupa case. Entire length 9 mm.; (*d*) imagines ♂ 6 mm., ♀ 4 mm. (Micro-slides, Waddington.) The eggs in gelatinous state show larvae emerging. Waddington. The larvae occur in wet places, ditches, etc.

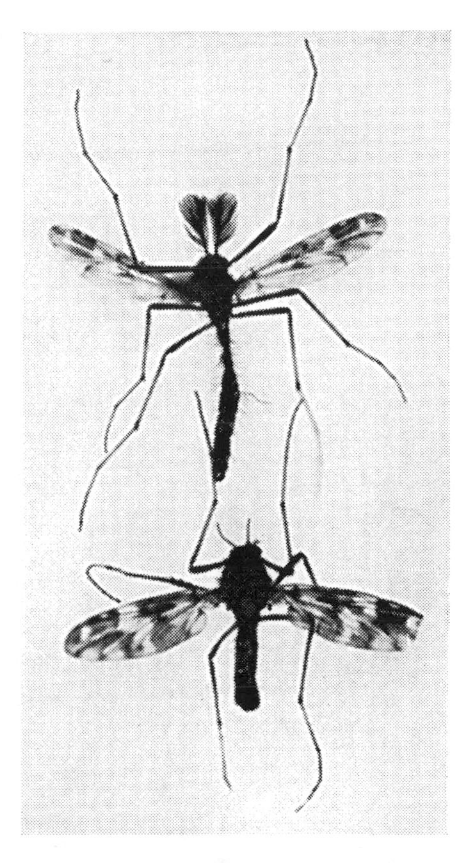

Fig. 4. *Tanypus varius* F. ♂ 6×8 mm. ♀ 4×11 mm. 18. iv. 19, N. D. F. Pearce, Grantchester. Common in Britain. Black with black and white splashed wings.

Fig. 5. *Corethra plumicornis* F. Ova, section of rush 6×1 mm. See *Typical Flies*, Series 2, figs. 11—17 (Micro-slide, Waddington.)

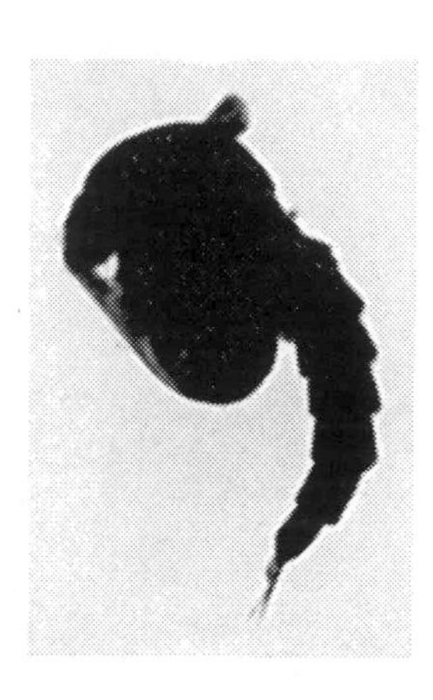

(*a*) (*b*)

Fig. 6. *Aedes cinereus* Mg. (*a*) Larva, $5\frac{1}{2}$ mm. (*b*) Pupa, $5\frac{1}{2}$ mm. Stroudens Farm, Bournemouth. (Micro-slides, Waddington.) See *Typical Flies*, Series 2, figs. 21, 22, Waddington. Dr Lang suggests "it is a river-haunting species."

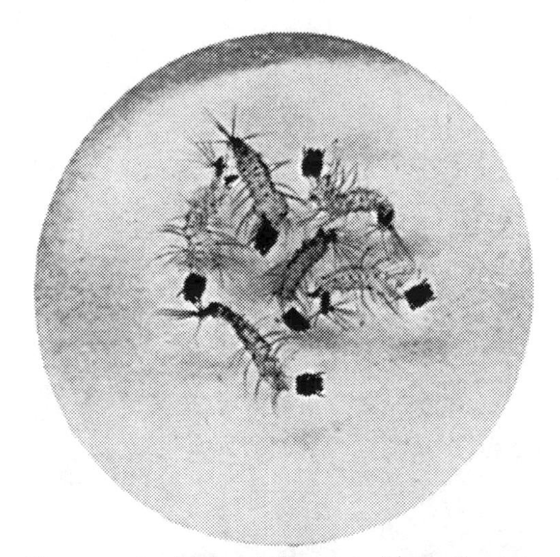

Fig. 7. *Anopheles bifurcatus* Linn. Larval skins, single skin $3\frac{1}{2}$ mm. Branksome, Bournemouth. 3. x. 19. (Micro-slide, Waddington.) See *Typical Flies*, Series 1, fig. 14; Series 2, fig. 23. Dr Lang states "Has been known to convey malaria in Italy." Sir Ronald Ross says: "for every million killed by malaria, 200,000,000 are rendered sick every year. Malaria kills more people than any other disease in the world." (This *Anopheles* strongly resembles *A. nigripes*.) The larvae of some *Anopheles* are found in hollowed trees.

(*a*)

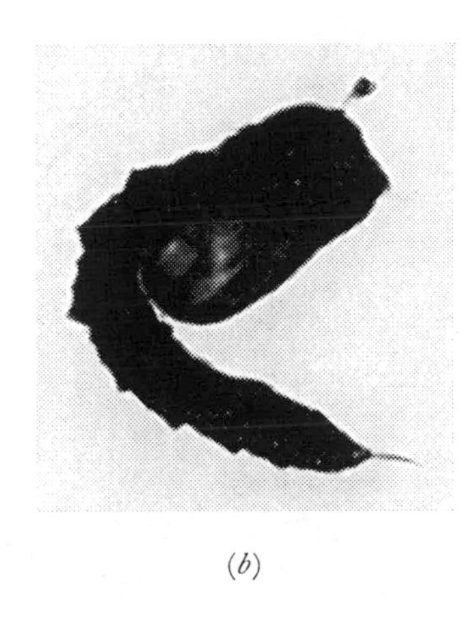

(*b*)

Fig. 8. *Dixa* Mg. (*a*) Larva, 4 mm. (*b*) Pupa, 3 mm. x. 12, Kempston, Bournemouth, fountain basin. (Micro-slide, Waddington.) See *Typical Flies*, Series 2, fig. 26. The larva is blackish on material in decay among reeds.

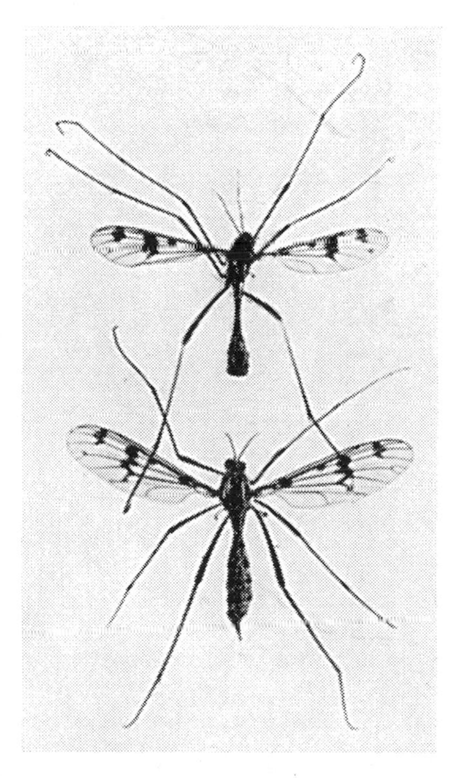

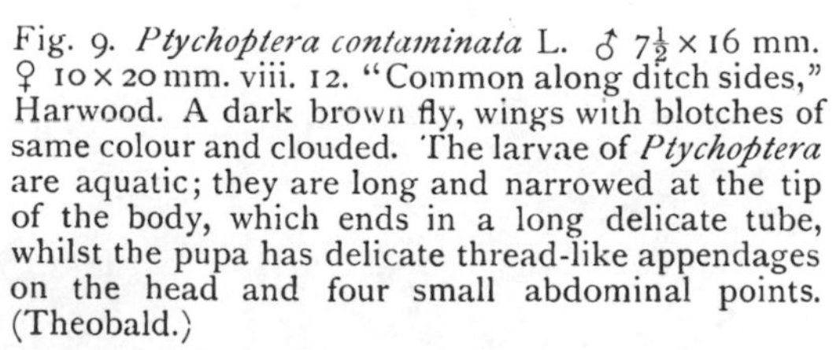

Fig. 9. *Ptychoptera contaminata* L. ♂ 7½ × 16 mm. ♀ 10 × 20 mm. viii. 12. "Common along ditch sides," Harwood. A dark brown fly, wings with blotches of same colour and clouded. The larvae of *Ptychoptera* are aquatic; they are long and narrowed at the tip of the body, which ends in a long delicate tube, whilst the pupa has delicate thread-like appendages on the head and four small abdominal points. (Theobald.)

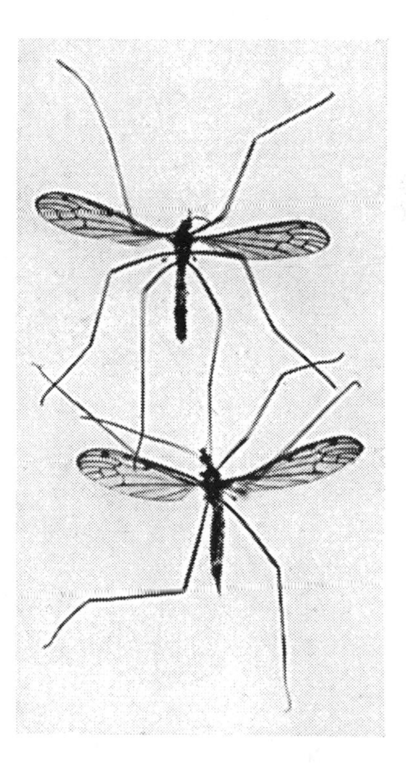

Fig. 10. *Limnobia nigropunctata* Schum. ♂ 9×22 mm. ♀ 11×22 mm. A brown fly, with brown veined wings with three brown spots on each; pale thorax. Taken by side of a small wood by sweeping. The larvae of this genus live in moist earth and fungi, some are aquatic; the body is supplied with numerous long filaments which contain air tubes. (Theobald.)

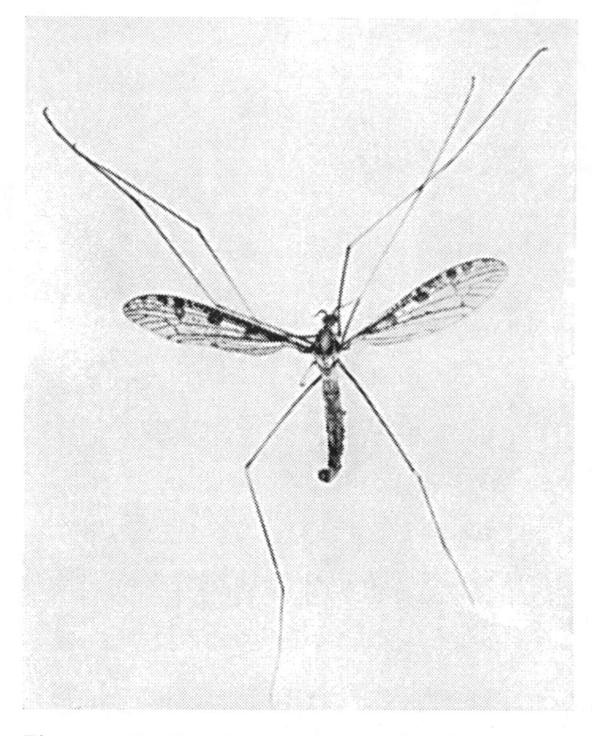

Fig. 11. *Poecilostola punctata* Schrk. ♂ 9×25 mm. A few by sweeping in damp meadows. Grey-brown abdomen, brown mottled wings and halteres.

Fig. 12. *Limnophila lineola* Mg. ♂ 11 × 24 mm. ♀ 12 × 29 mm. v. 12. A pale brown fly with yellowish shading and wings of same colouring.

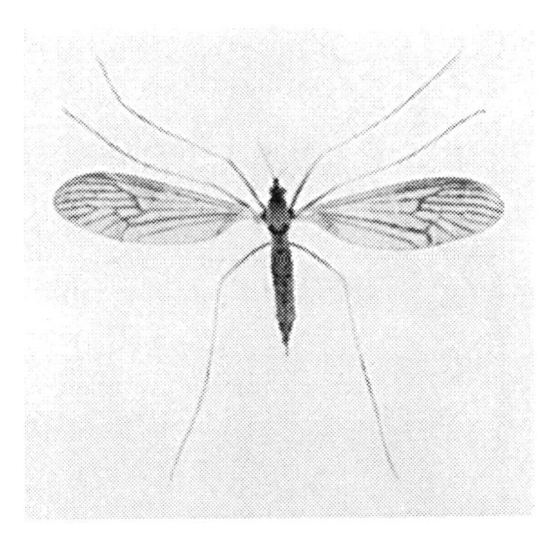

Fig. 13. *Trichocera hiemalis* Deg. ♀ 5 × 14 mm. 16. x. 12. Horkesley. Dark brown and brown veined wings. "Winter Gnat." Uncertain how many British species of the genus. The larvae live in decaying wood and roots, especially potatoes, turnips and carrots. They also occur in fungi; although not actually destructive, they hasten the decay of roots. (Theobald.)

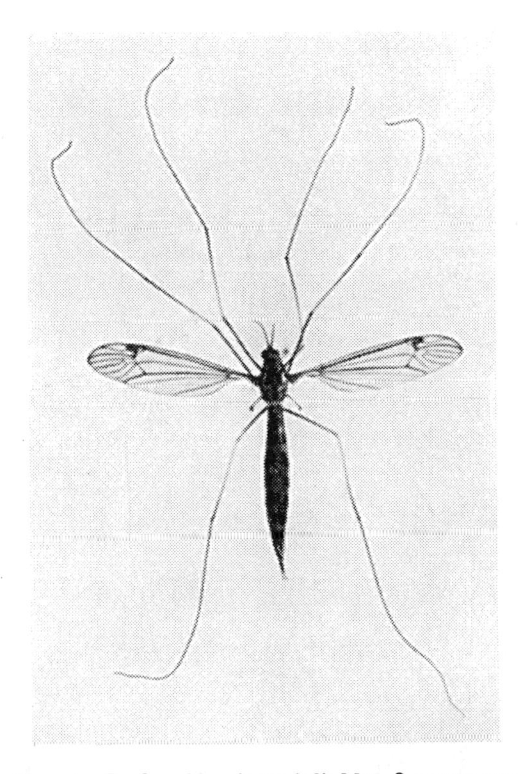

Fig. 14. *Pachyrrhina imperialis* Mg. ♀ 17 × 31 mm. Abdomen orange with broad black markings; thorax of paler yellow colouring striped or banded black; legs brown.

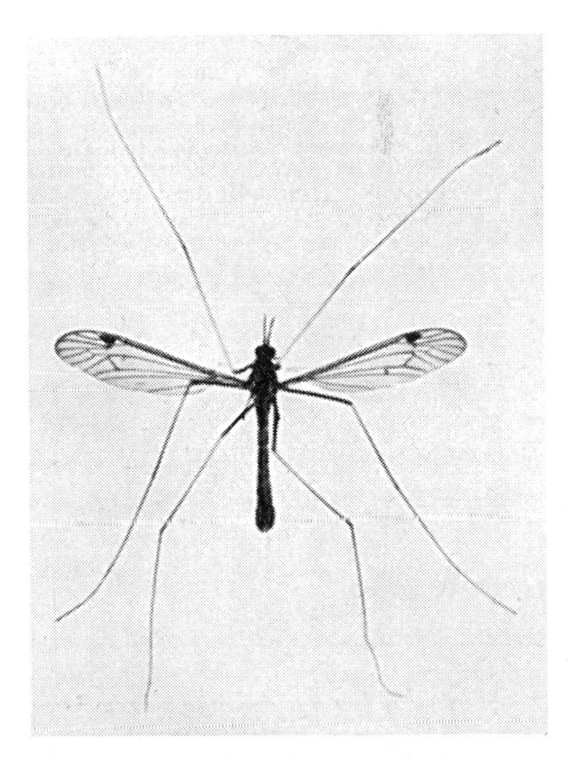

Fig. 15. *Pachyrrhina guestfalica* Westh. ♂ 10 × 22 mm. A pale brown fly with a single brown spot on brownish wings.

Fig. 16. *Tipula confusa* v. d. Wulp. ♂ 10 × 28 mm. ♀ 11½ × 26 mm. On windows of house, Blairgowrie, ix. 21. Found commonly breeding on moss on the Roman Wall, Colchester, September and October. (Harwood.) Dark brown with marbled wings.

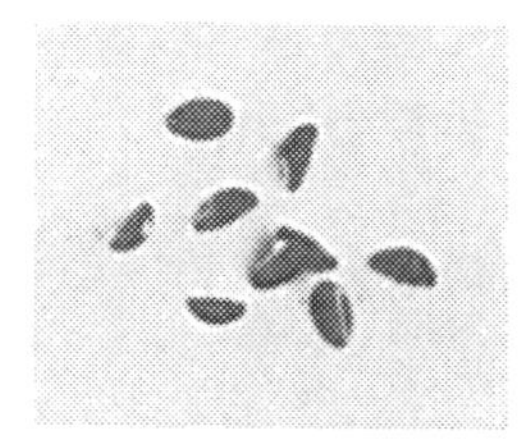

Fig. 17. *Tipula vernalis* Mg. Ova (each) about 1 mm. See *Typical Flies*, Series 1, fig. 23. (Harwood.)

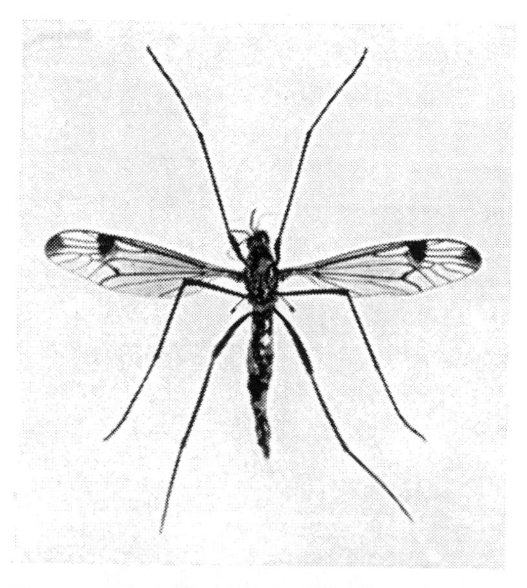

Fig. 18. *Dictenidia bimaculata* L. ♀ 12½ × 22 mm. Scarce. Once found in some numbers on Alder bushes near Alresford, Essex. A black fly with orange-brown legs and brownish wings.

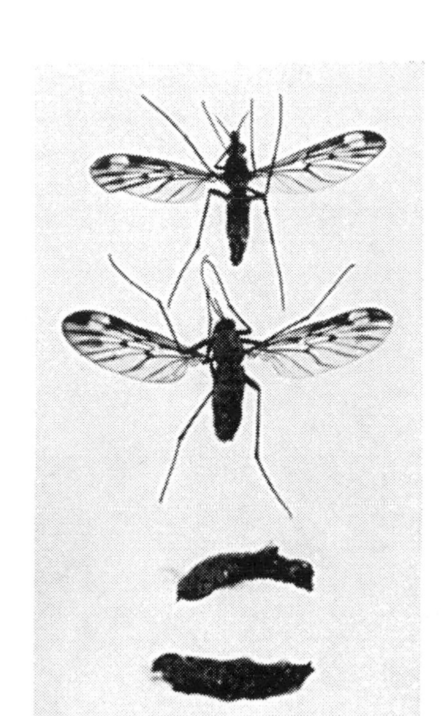

Fig. 19. *Rhyphus fenestralis* Scop. ♂ 5 × 11 mm. ♀ 6 × 12 mm. Pupae 7 mm. Found in rotten excavated potatoes in large numbers; the grubs pupated in the potato. Harwood. Also found in over-ripe fruit, and dung; and have been recorded from bee comb in diseased hives. The larva is cylindrical and very long, with four small abdominal tubercles. (Theobald.) ♂ with dark abdomen, ♀ buff abdomen, brown bands, mottled wings, brown legs.

♂

♀

Fig. 20. *Pachygaster atra* Pz. ♂ $3\frac{1}{2} \times 7$ mm. ♀ $4 \times 8\frac{1}{2}$ mm. On ivy at Grantchester, N. D. F. P. A small black fly, with blackish wing shading; the larvae live in decaying wood.

Fig. 21. *Stratiomys potamida* Mg. ♂ 14×28 mm. ♀ 13×26 mm. Larvae aquatic. These flies appear on or near plants bordering rivers. Rhinefields. New Forest, Adams, "not common." The ova are laid under the leaves of water plants and are arranged like tiles on a roof. (Theobald.) ♂ dark and abdomen thickly covered with black velvety bands, three splashes of orange at sides of abdomen, orange legs. ♀ back of head edged yellow, two deep black bands on yellowish abdomen, semi-circular black band next thorax. A very handsome fly.

Fig. 22. *Stratiomys furcata* F. ♂ 13½×24½ mm. ♀ 15×26 mm. On Umbellifera on the East Coast, 1898 numerous (generally scarce). ♂ dark head, eyes hairy (Verrall) and body, sides of abdomen splashed yellow-orange, legs black and orange. ♀ dark, abdomen marked vividly at sides and end with light orange splashes; head of female with orange streak.

Fig. 23. *Stratiomys longicornis* Scop. ♂ 13×22 mm. ♀ 13×26 mm. Occasionally, both inland and on the coast. A dark brown fly and pale thorax in both sexes. Walker says the eggs of *S. Chamaeleon* are on the underside of leaves of *Alisma Plantago*, the water plantain; the eggs from white become green and are arranged as tiles on a roof.

Fig. 24. *Odontomyia argentata* F. ♂ 9×17 mm. ♀ 8×18½ mm. Scarce. Both thorax and abdomen are broader in the ♀ and stripes more conspicuous, being pubescent and golden. The ♂ is darker and more silvery. By sweeping and searching, April 25—May 10, 1921. (Harwood.) The ♂♂ appear some days before the ♀♀, Henny, Essex.

Fig. 25. *Odontomyia tigrina* F. ♂ 11 × 15 mm. Larval skin 15 × 3 mm. Pupa develops inside larval skin. A very dark fly, base of wings darkened.

Fig. 26. *Haematopota crassicornis* Whlbg. ♂ 10 × 20 mm. ♀ 9 × 18½ mm. Nethy Bridge, vii. 13, P. Harwood, Wareham and Morden Common, Dorset. The larvae of Tabanidae found in damp soil and on decaying animal and vegetable matter, and also on live earthworms, etc. (Theobald.) Darker in wings than *pluvialis*, abdomen lighter marks. (Not common: Adams.)

(*a*)

(*b*)

Fig. 27. *Therioplectes tropicus* Mg. (*a*) ♂ 13×26 mm. ♀ 14×21 mm. (Determined by E. E. Austen.) Taken by H. Jones, New Forest, 10. vi. 22. *Therioplectes* "eyes pubescent," Verrall. ♀ thorax black, abdomen black, buff ochrish mark on side of second section. Black hairs on side of thorax, central portion of legs lighter than ♂. (*b*) ♂ further enlarged to show head. Black thorax, centre of abdomen with black streak running down it, sides ochrish brown on two sections, legs dark brown, wings brown veined. "The ♀♀ only of the Tabanidae bite. ♂ Tabanidae are found in certain seasons, over water, the New Forest being specially favoured. *Therioplectes* is more sluggish in its habits than *Tabanus*, often resting on overhanging boughs or herbage (invariably, in my experience, after drinking), whilst *Tabanus* comes straight down from a height, just sips the water and flies up again making a capture exceedingly difficult. The ♂ *Atylotus*, on the few occasions observed, seems to possess a mixture of the habits over water, of both the above. In flight (I have not yet seen it settle) it is rather difficult to distinguish from a bright coloured ♂ *distinguendus*." (Note contributed by H. Jones.) Professor Hine, U.S.A., speaks in his pamphlet (reprint, mentioned in Preface) of the Tabanidae ovipositing on leaves of water plants; and their collecting and dispersing at early morning, on certain trees and palings, for pairing; (the eggs black) under much the same circumstances of capture as in the New Forest.

Fig. 28. *Tabanus solstitialis* Schin. Br. Meigen (?) form *distinguendus* Verrall. ♂ 15×29 mm. (Determined by E. E. Austen.) A series taken by H. Jones shows very divergent colouring. ♂ thorax dark, abdomen decided orange to brown (with darker tinged greyish hairs) and central streak. A series of six ♂s shows special great divergence in colouring. "The typical race of *T. solstitialis* is far less common than the form *distinguendus*" (E. E. Austen).

Fig. 29. *Atylotus fulvus* Mg. ♂ 15×28 mm. ♀ 16×28 mm. Park Hill, New Forest, H. Jones. Wareham Heath, 1917–1918, E. K. P. A dull orange fly, with hind legs orange, fore legs partly brown, exceedingly golden in life, ♂ thorax slightly darker than ♀, eyes brilliant green, a most beautiful fly, losing its colouring in death.

Fig. 30. *Atylotus latistriatus* Brauer. ♂ 14×27 mm. Golden brown with central mark of much darker brown shade. ♀ 13½×26½ mm. Two dark bars length of abdomen, which is ochrish in shade. Very rare, on Essex coast, occasionally on flowers of Sea Holly. (Harwood.)

Fig. 31. *Tabanus bovinus* L. ♂ 22 × 42 mm. H. Jones, New Forest, vii. 1921. For ♀ see Series 1, fig. 42. "*Tabanus* eyes bare," Verrall. Thorax dark and striped, brown stripes on abdomen relieved by orange stripes growing gradually narrower, fore legs darker than hind pair which are chiefly orange to orange-brown. Six were taken over water by H. Jones, all ♂♂, vii. 1921.

Fig. 32. *Tabanus autumnalis* L. ♂ 16 × 36 mm. ♀ 18 × 39 mm. (Determined by E. E. Austen.) ♂ has yellow orange markings on the abdomen, shading to brown at apex, dark thorax. Harwood. ♀ grey striped thorax and alternately grey and dark slanting patches on abdomen.

Fig. 33. *Tabanus autumnalis* L. ♀ 18 × 39 mm. Pupa case 24 × 5 mm., dark biscuit colour, bred Harwood, 1919. On tree trunks and posts. "I have never known this species to bite" (Harwood). Dark streaks on grey thorax, three grey marks separated by dark splashes on the sections of abdomen, extremities of legs dark, centres ochrish.

Fig. 34. *Tabanus bromius* L. ♂ $15\frac{1}{2}$ × 27 mm. (8. vi. 21, Jones.) ♀ 16 × 28 mm. (vi. 15, Silchester, Harwood.) (Determined by E. E. Austen.) ♂ greyish thorax, grey pubescence, abdomen with alternate slanting dark and light buff stripes on all sections), legs ochrish to dark brown at extremities. ♀ very like ♂ in colouring, but thorax more striped, and legs flatter and wider.

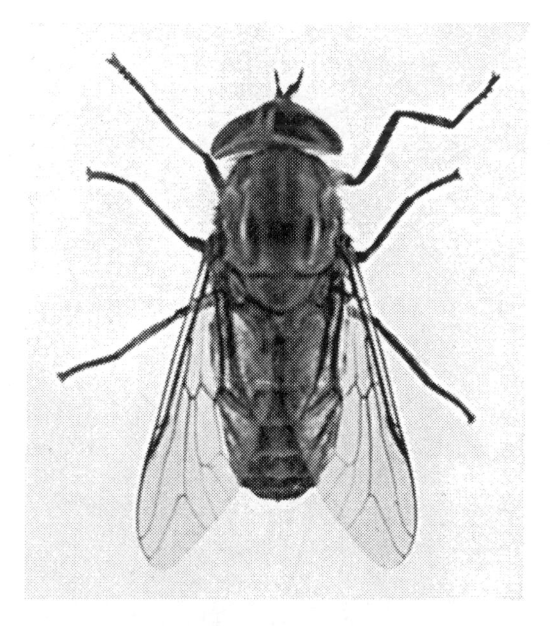

Fig. 35. *Tabanus bromius* L. ♀ 14 mm., folded wings. Holmsley, New Forest.

Fig. 36. *Tabanus maculicornis* Ztt. ♂ 14 × 26 mm. ♀ 13 × 25 mm. (Determined by E. E. Austen.) ♀ grey thorax, grey flecks on dark abdomen, legs particolour dark brown and ochrish; brown veined wings.

Fig. 37. *Tabanus maculicornis* Ztt. Two ♂♂ to compare sizes: 14 × 26 mm., 10 × 21 mm. 10. vi. 22, H. Jones, New Forest. (Determined by E. E. Austen.) "The smallest species of the genera," Verrall. Black thorax fringed grey hairs. Sometimes chestnut splashes on two upper sections of abdomen, grey narrow line between sections of abdomen.

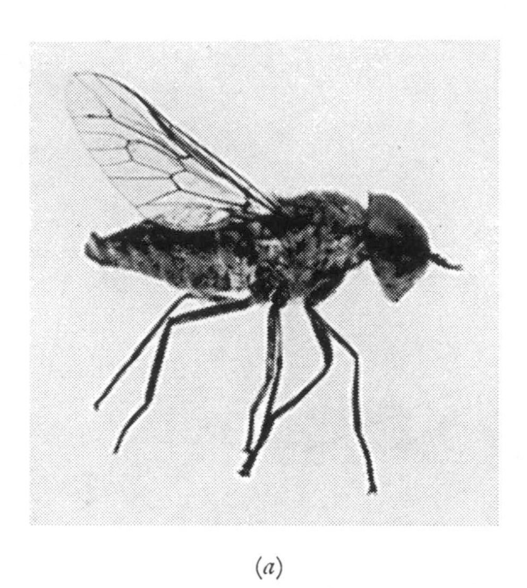

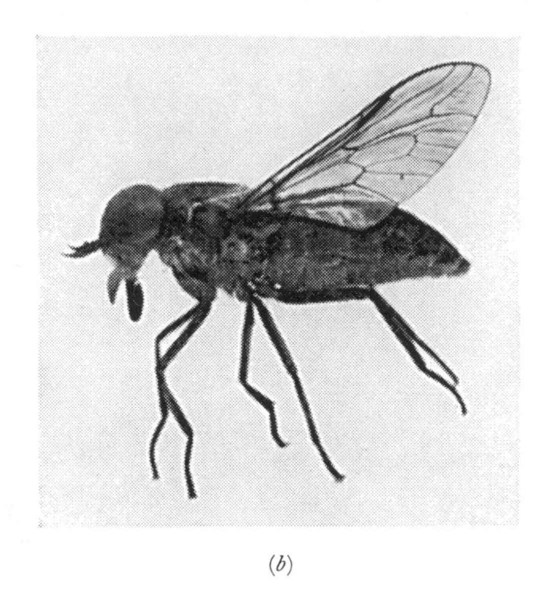

(*a*) (*b*)

Fig. 38. *Tabanus maculicornis* Ztt. (*a*) ♂ 14 mm. (*b*) ♀ 12½ mm. To show mouth parts. Jones, New Forest.

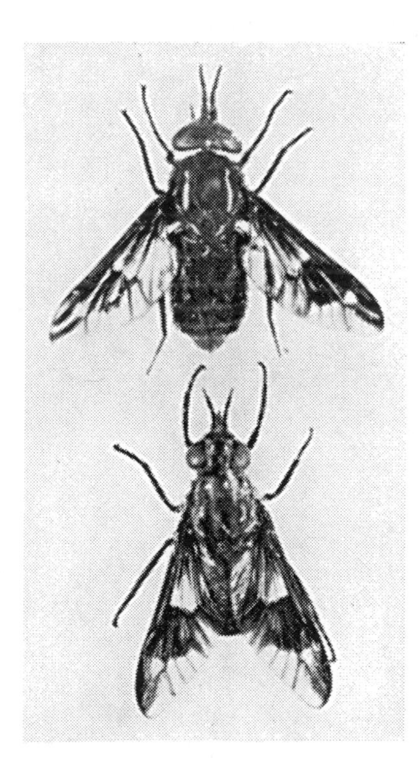

Fig. 39. *Chrysops relicta* Mg. ♂ frons to extremity of abdomen 9 mm. ♀ frons to extremity 8 mm. Both sexes by sweeping nettles (as with *C. caecutiens* the ♂ much the rarer). Abdomen yellow with black bands and triangles, dark black clouded wings; ♂ darker than ♀, yellowish pubescence, black antennae, eyes green, with spots of very small size (in *C. sepulcralis* the eyes are emerald green and in certain lights after capture seem to turn golden red). The following notes on *Chrysops* have been given me by Major E. E. Austen, D.S.O., F.R.S., British Museum: "The worm causing Calabar swellings in West Africa is now known as *Loa loa*: and the species was described by Guizot in 1778. The metamorphosis of the embryos of the worm in the two species of *Chrysops* (*C. dimidiata*, v. d. Wulp, and *C. silacea*, Austen) was discovered by Dr R. Leiper, in 1912, Leiper thus proving what had originally been suggested by Manson some 20 years earlier. According to Dr and Mrs Connal, who, in southern Nigeria in 1921, amplified and completed Leiper's work, the metamorphosis of the *Loa loa* embryos takes place, not in the salivary glands of the fly, but in its muscular and connective tissue—chiefly in the abdomen. When metamorphosis is complete, the larval worms assemble in the fly's head, at the base of the proboscis."

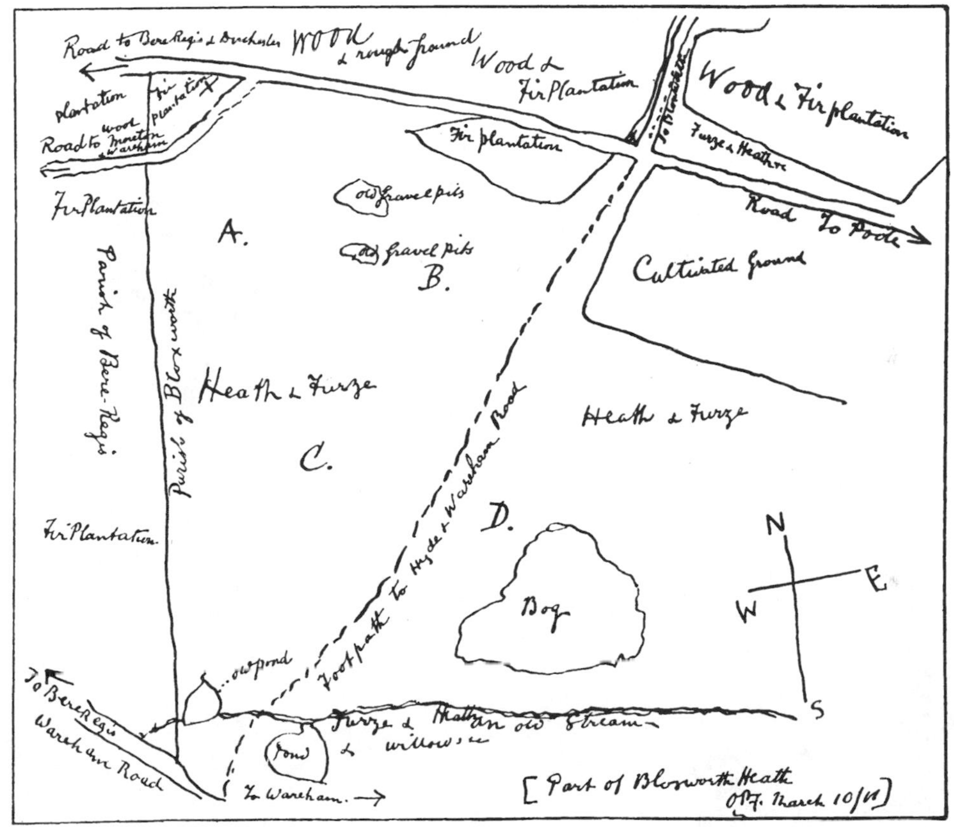

Fig. 39 *b.* An additional map of finds of *Chrysops sepulcralis*; by the late Rev. O. Pickard-Cambridge, Bloxworth, Dorset. "*A, B, C, D* found *Anthrax circumdatus* on bare and arid spaces. X represents, on the left-hand, route to Wool Station: on right-hand, to Bloxworth. *Chrysops sepulcralis* also found here (as certified by Verrall); many trees marked have disappeared." In 1904 "he (Mr Cambridge) estimated he had taken 20 *C. sepulcralis* in the last 30—40 years," says Verrall and this was confirmed to me by the above in 1911 when the map was given me by him.

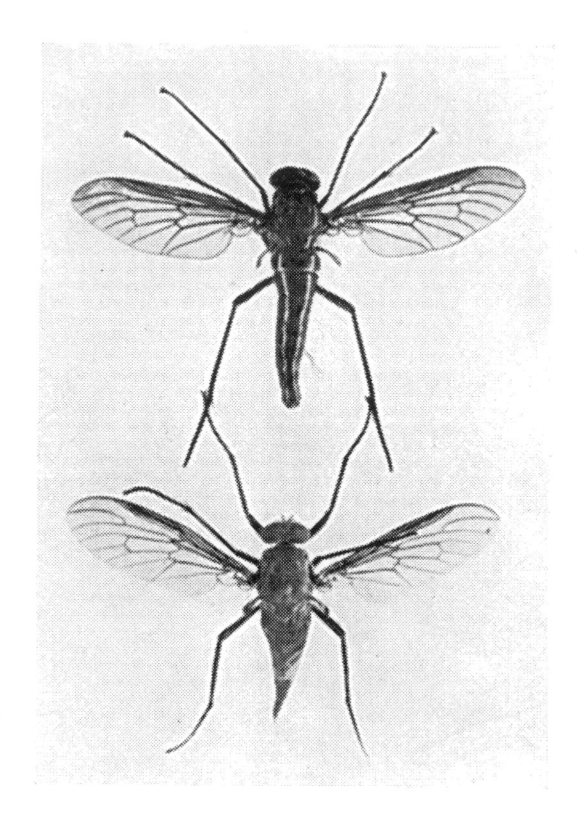

Fig. 40. *Leptis tringaria* L. ♂ 11 × 21 mm. ♀ 11 × 20 mm. New Forest, 15. vi. Abdomen orange with black spots. Said to sit on trees with head downward. "The larvae of Leptidae live in the earth—are carnivorous and devour other larvae, or penetrate into beetles which they devour; they also feed on earthworms." Lundbeck (as stated by Marchal).

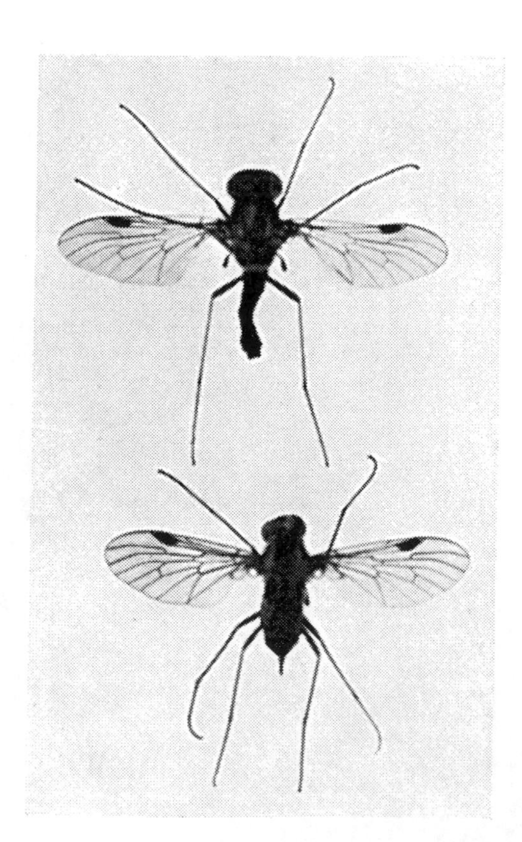

Fig. 41. *Chrysopilus cristatus* F. ♂ 8 × 16 mm. ♀ 6 × 15 mm. N. D. F. P., Grantchester. Abundant in damp meadows. A brown fly, palish legs, and brown tinted wings.

Fig. 42. *Atherix Ibis* F. 8 × 19 mm. Wicken Fen, 29. vii. 18, N. D. F. P. Clear wings with brown spots, abdomen somewhat tawny.

Fig. 43. Cluster of *Atherix Ibis* on Sallow. $2\frac{1}{2} \times 2\frac{1}{2}$ inches and about 1 inch thick actual size. Found by Dr R. C. L. Perkins, South Devon, 1917. The females oviposit in common, on a branch overhanging water, from which the larvae drop to the water below. The flies dying after ovipositing, add their bodies to the mass. Several hundreds in number. (The cluster had a strong odour on arrival.) Curtis, Nat. Hist., vi. pp. 480. See F. Walker, Vol. I. p. 70, Insecta Britannica, *Diptera*. Now in Cambridge University Museum. Photo of cluster: T. Pearce, by permission.

Fig. 44. *Lasiopogon cinctus* F. ♂ 10 × 16½ mm. ♀ 10 × 17 mm. Adams, New Forest. A black fly, with narrow white bands on the abdomen and grey striped thorax.

Fig. 45. *Rhadiurgus variabilis* Ztt. ♀ 13½ × 20½ mm. vii. 13, Nethy Bridge, P. Harwood. A dark brown fly, with bristled brown legs, lighter brown wings. The food of the Asilidae consists of other insects, and they may be seen sitting on hot sands awaiting their opportunity; some are even cannibalistic. The larvae live in rotting bark. *Dioctria* also prey on small insects, and are often found with them.

Fig. 46. *Epitriptus cingulatus* F. ♂ $10\frac{1}{2} \times 16$ mm. ♀ $13 \times 17\frac{1}{2}$ mm. ♂ a pale brown smooth bodied fly. ♀ dark brown, smooth bodied, legs and claws densely haired. On the sandy wastes, W. Suffolk, also Royston Heath, Herts. (Harwood.)

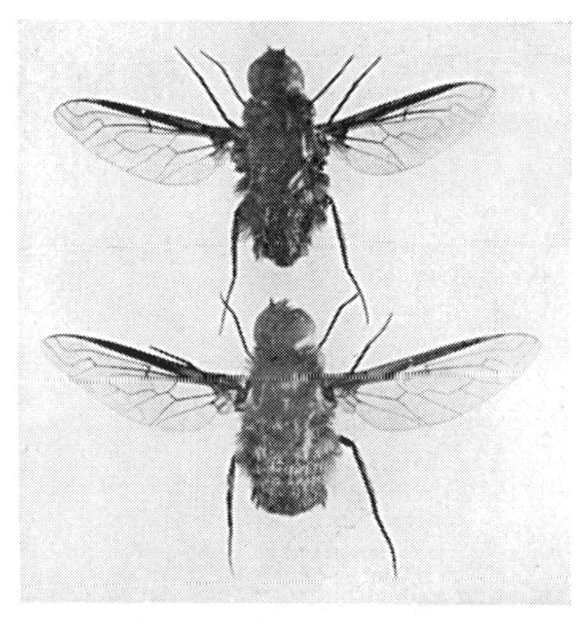

Fig. 47. *Villa* (*Anthrax*) *paniscus* Rossi. ♂ 12×26 mm. ♀ 12×28 mm. On coast sands, not rare, and Studland, Dorset. Anthrax larvae are parasitic on Bees (Megachile, Osmia, Odynerus) and some Noctuids. (Theobald.) ♂ brown abdomen, light pubescence, of a golden type; ♀ slightly more coloured abdomen.

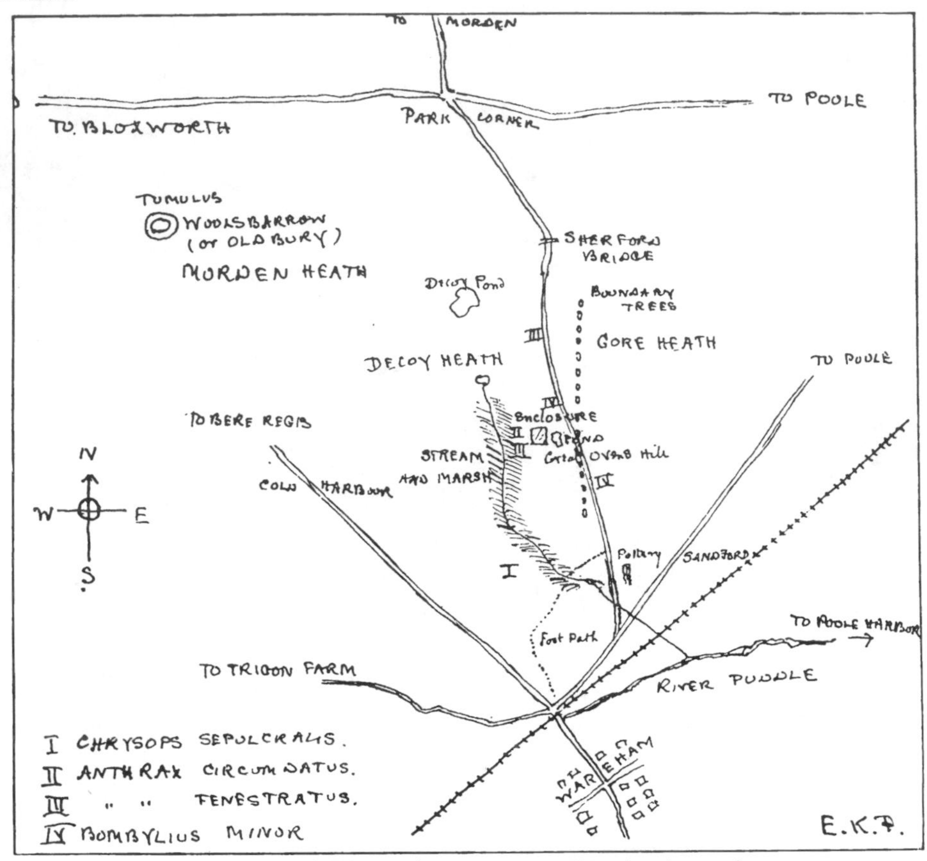

Fig. 48. Map showing areas of *Diptera* (in Dorset, near Wareham); as given on its margin, these are generally found on special oases; with wide heath areas destitute of specimens. I. also found, to right of bog, vii. viii. 1917, 1918.

Fig. 49. Old coaching milestone near Wareham, a noted bog, same district, marking the whereabouts of the stream, and reed bog, by roadside.

Fig. 50. Wareham Heath Bog. The habitat of *Chrysops*, *Anthrax*, *Bombylius*. See *Typical Flies*, Series 2. Williston (U.S.A.) mentions that the larvae of *Anthrax* have been found parasitic upon *Megachile*, *Odynerus*, etc. (the arid heath regions of Dorset are so populated. E.K.P.).

Fig. 51. The Path between the bog; a causeway where 140 *Chrysops sepulcralis* congregated on verge of Morden Heath, vi. 1917—18. A solitary region plentiful in "Dragon flies," solitary bees and wasps.

Fig. 52. *Bombylius major* L. 11 × 38 mm., amber pubescence of body, very decidedly marked brown fore part of wings. Pupa 11 × 5 mm. Found 20. iv. 21 by Miss Harwood under a stone in a garden at Sudbury, Suffolk (being biscuit colour). Bred out 21. v. 21, Harwood; a few found in Spring annually at Grantchester by N. D. F. P. Very partial to Primrose and Bugle blossom. The Bombylidae are known to be parasitic upon *Halictus* and other solitary bees. The larvae probably feed on larvae of the bees.

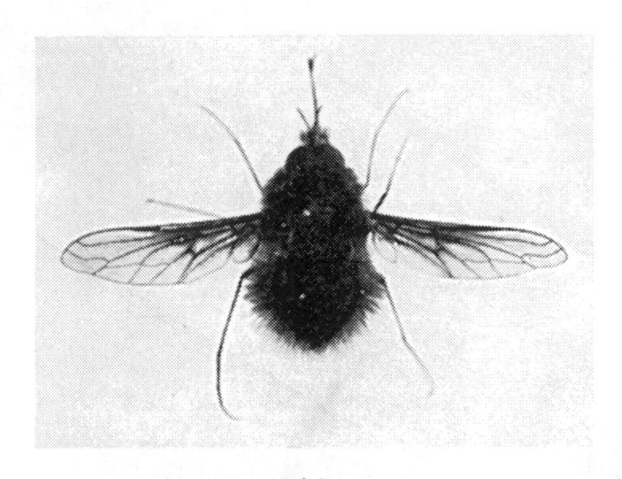

(a) (b)

Fig. 53. *Bombylius canescens* Mik. (a) ♂ 7 × 18½ mm. (b) ♀ 8 × 18 mm. Distinguished from *B. minor* by a postocular fringe of black hairs, see *Typical Flies*, Series 2, figs. 49, 50. Taken by the late A. E. J. Carter, Blairgowrie, Perthshire, 10. vi. 10. The larvae of *Bombylius* are found parasitic on larvae of Bees of the genus Andrena, Colletes. (Lundbeck and Verrall.)

Fig. 54. *Thereva nobilitata* F. ♂ 10 × 18 mm. ♀ 10½ × 17 mm. Window, Kempston, Bournemouth, 1. viii. 22. Probably the commonest of the genus. ♂ brown with light pubescence of abdomen. ♀ paler tawny grey.

Fig. 55. *Thereva annulata* F. ♀ 10 × 16. Clacton-on-Sea, vii. 1911, and at Haven, Sandbanks, Poole and Matley Bog, New Forest, Adams. Ashy grey abdomen with black stripes. "*T. plebeia* larvae have been found injuring beet." (Theobald.) They are vermiform, with pointed extremities. The larvae of Therevidae live in the soil, pale and vermiform in shape, sometimes causing damage to potatoes. (Theobald.) Adult Therevidae prey on other diptera.

Figs. 56, 57. The haunts of the Therevidae. Sandbanks, Poole Haven, where these flies crawl swiftly, watching for prey, beneath the reed grass—*Ammophila arundinacea*, Sea Reed or Marram, which was formerly protected by Acts of Parliament, because of its binding qualities on Sea Coast sands.

♂

♀

Fig. 58. *Oncodes gibbosus* L. ♂ 6 × 11 mm., bone coloured with black bands on abdomen. ♀ 5 × 13 mm., dark brown with light lines on abdomen. Parasitic on Spiders, which are entirely eaten away by the larvae of *Oncodes*, which larvae are said to be able to leap. Found by sweeping warm heather clad slopes, under hawthorns: Jones, N. Dorset. The late Prof. Williston in *N. American Diptera* says, cobwebs of common spiders contained empty spider skins, and soft white maggots with nearly spherical bodies. Lundbeck says its eggs are found on dry branches near rocks. Piffard found them at Brockenhurst, New Forest Railway bank: very local.

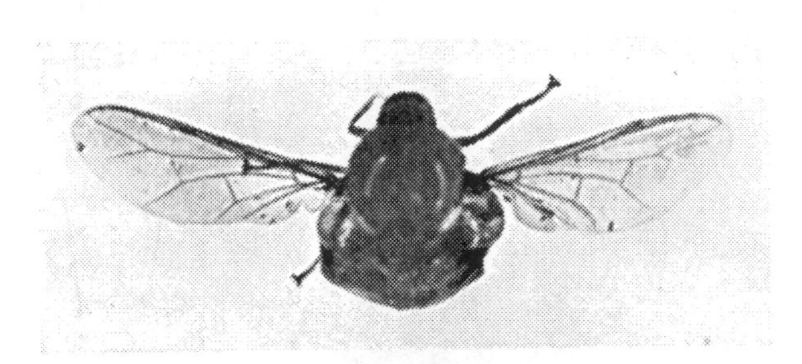

Fig. 59. *Acrocera globulus* Pz. ♀ 4 × 11 mm. Harwood. Verrall states that the larvae are parasitic on spiders (*Tegenaria agilis*, etc.) and distinguishes it from *Oncodes* by the vertical position of the antennae and by the venation. "Found at Ower Quay, Poole, gregarious." (Haines.) A pale brown fly, wings very transparent but more marked than in *Oncodes*. "The species of this family generally, only found now and then." (Lundbeck.) Glanvilles Wootton (Dale): Corfe Castle (reported).

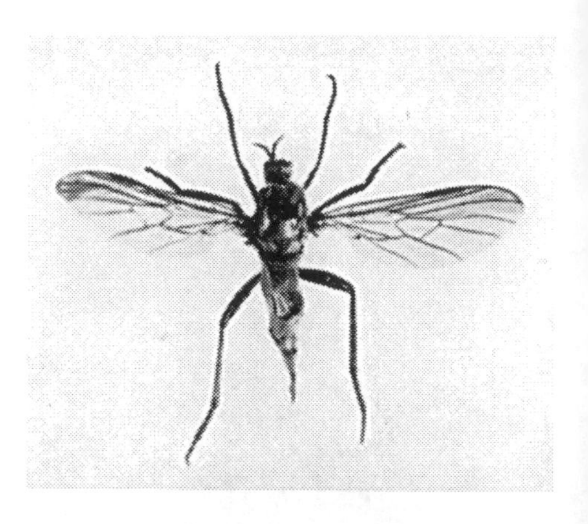

Fig. 60. *Rhamphomyia sulcata* Fln. ♀ 8 × 14 mm. "Plentiful in New Forest." (Adams.) A dark brown fly, dark legs same tint, brownish wings.

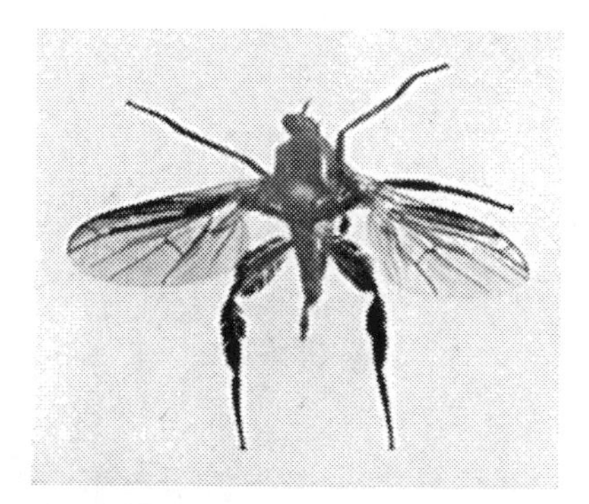

Fig. 61. *Empis pennipes* L. ♀ 3 × 6½ mm. ♀ "Hind femora sometimes feathered." (Walker.) From Grantchester, N. D. F. P. "The Empidae are predaceous and found in damp regions usually." (Theobald.) A black fly, shiny body and feathered hind legs.

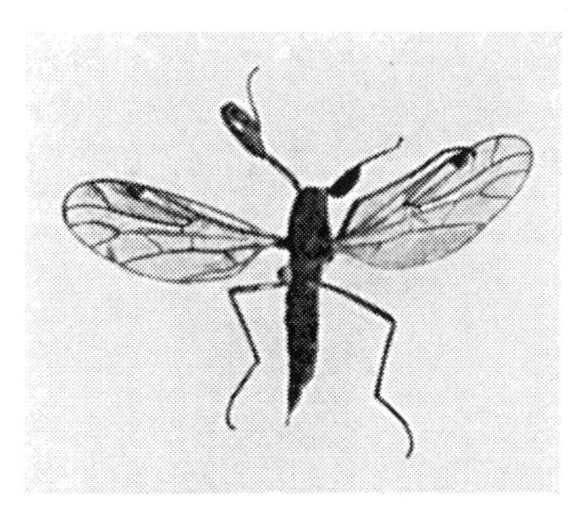

Fig. 62. *Hemerodromia precatoria* Fln. ♀ 5 × 10 mm. They prey on minute *Diptera* and other insects; frequent banks of rivers and lakes, on grassy spots. (Walker.) A yellow fly with broad black bands on abdomen and curiously dilated yellow fore legs.

Fig. 63. *Poecilobothrus nobilitatus* L. ♂ 8 × 14 mm. A bronze fly, bronze legs. Grantchester, N. D. F. P. Professor Aldrich in Williston's *N. American Diptera* notes "the sexual ornaments of the Dolichopodidae; paraded in their case as by Peacocks and ♂ Turkeys."

Fig. 64. *Machaerium maritimae* Hal. ♂ 5 × 11 mm. ♀ 6½ × 13 mm. A greenish fly with hyaline wings. The ♂ has the fifth segment of the abdomen turned in. Common on marshes near the sea.

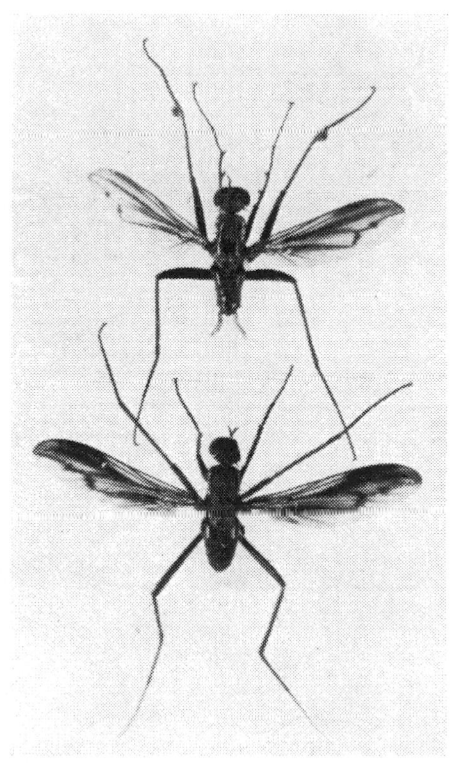

Fig. 65. *Scellus notatus* F. ♂ 6 × 12 mm. ♀ 4 × 15 mm. Clacton, vii. 12, Harwood. A dark bronze fly, wings veined in dark brown and darkly tinted throughout.

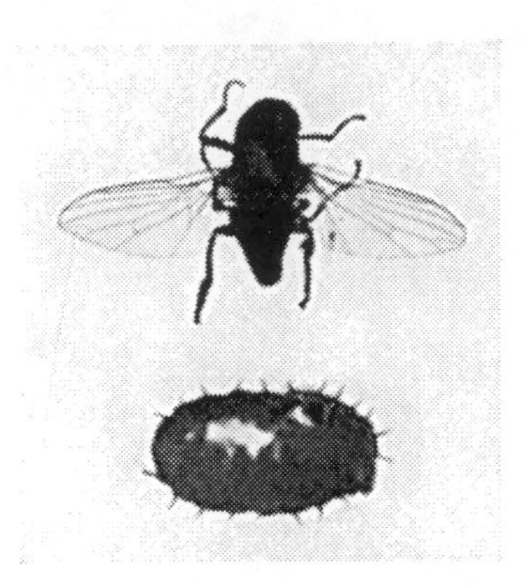

Fig. 66. *Platypeza dorsalis* Mg. 3×6 mm. Pupa $3\frac{1}{2}$×2 mm. From *Agaricus campestris* (E. K. P.), Morden, Dorset. Bred out N.D.F.P. Larvae live in fungi. A black fly, pupa dark brown.

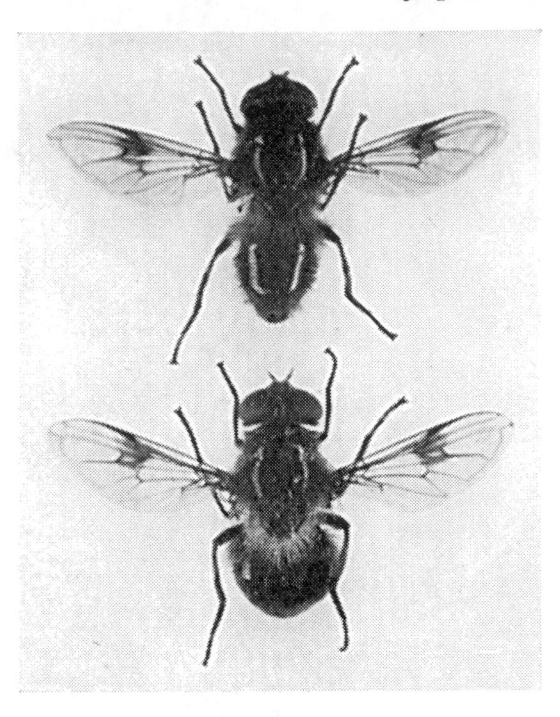

Fig. 67. *Chilosia illustrata* Harr. ♂ 12×24 mm. ♀ 12×24 mm., abdomen wider. "Not common," Adams, New Forest. Thorax of ♀ fringed grey hairs, abdomen black, tipped golden pubescence; abdomen of ♀ wider than ♂.

Fig. 68. *Chilosia grossa* Fln. ♂ 13×25 mm. ♀ 11×$24\frac{1}{2}$ mm. "Not common," Adams, New Forest. On Sallow blossom in April. A bronze fly, sides of abdomen fringed bronze hairs. Ford mentions "bred from *Carduus palustris*," 2. iii. 8; damp meadows.

Fig. 69. *Sphaerophoria scripta* L. ♂ 11 × 15 mm. ♀ 8×12 mm. Transparent wings, abdomen orange with black bars or bands.

Fig. 70. *Sphaerophoria menthastri* L. 8 mm.: cocoon 7 mm. Shown as emerged. The larvae of Syrphidae feed on Aphids. Abdomen pale orange, brown bands; pupa case grey white.

Fig. 71. *Brachyopa bicolor* Fln. ♂ 8½ × 18 mm., 30. vi. 13. "Took a good many at a cossus infected oak, 1919," Adams. "Not to be found every year." The late Col. Yerbury found it in Boldre Wood; Adams, at Clay Hill Lodge, Fox Lease Park. Black thorax, chestnut brown to yellow abdomen, brown veined wings.

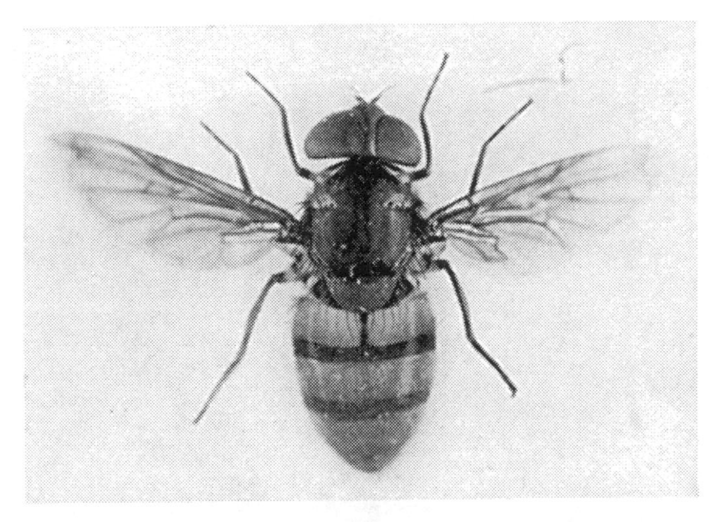

Fig. 72. *Volucella inanis* L. 16 x 28 mm. Jones, New Forest, 1921. Larvae are scavengers in Hornets' nests (*Vespa crabro*). Colouring of fly is yellow with black bands. Clouded wings.

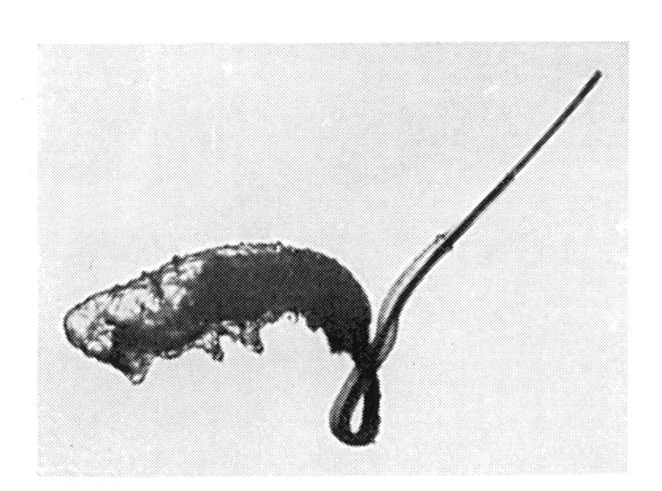

Fig. 73. *Eristalis tenax* L. Larva 8 mm., extending telescopic tube (tail) 13 mm. through which it breathes. "The Rat-tailed Maggot" lives in stagnant and filthy water, even liquid manure. (Micro-slide, Waddington.) See *Typical Flies*, Series 1, fig. 108.

Fig. 74. *Eristalis tenax* L. Pupa skin, 12 mm. (to end of tube 15 mm.); emerged fly, ♀ 15 mm.

Fig. 75. *Myiatropa florea* L. (P. Harwood.) ♀ 13×26 mm. Pupa 13½×6 mm. Brown. Orange abdomen with black markings, thorax of darker hue, clear wings.

Fig. 76. *Cynorrhina fallax* L. ♂ 10×19 mm. Nethy Bridge, vii. 13. P. Harwood. Mentioned by Verrall as found in this locality. Black fly but orange tip to abdomen conspicuous.

Fig. 77. *Microdon mutabilis* L. ♂ 11×18 mm. 15. v. 21. Jones, Roydon, New Forest. Probably from nest of *Formica fusca.* The larva (from nest of *Lasius fuliginosus*) 7×5½ mm.; note its small head on underside. From Aldridge Hill enclosure, 15. iv. 20. These curious larvae are only found in ants' nests, under bark, or stumps of trees, *i.e.* old stumps. The grub seems to rove through the passages of the nest made of chewed bark and decayed wood. (By early naturalists these larvae were thought to be molluscs.) See *Typical Flies*, Series 2, figs. 123, 124. Jones, New Forest.

Fig. 78. *Conops vesicularis* L. ♂ 13×24 mm. ♀ 14½×23 mm. 30. v. 15. *Conops* are flower-loving flies, parasitic on bees and wasps (said to oviposit on them in flight), and greatly resembling them. The larvae occupy the host's entire body. ♂ darkly infumated wings, dark brown abdomen with orange band. ♀ less darkly marked wings, yellowish abdomen. Adams.

Fig. 79. *Conops quadrifasciata* Deg. ♂ 11 × 17½ mm. ♀ 10×16 mm. (Harwood.) ♂ orange abdomen banded black, orange tip to abdomen. ♀ pear shaped abdomen, legs orange, same colour as abdomen (banded black).

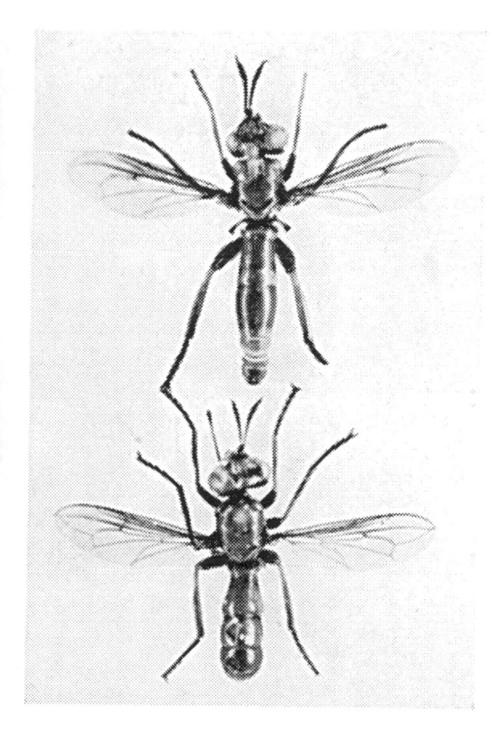

Fig. 80. *Conops ceriiformis* Mg. ♂ 13 × 19 mm. ♀ 11 × 18 mm. (Harwood.) ♂ black abdomen, narrow orange bands. ♀ orange abdomen, dark brown thorax. The forms of Conopidae vary somewhat.

Fig. 81. Habitat of *Conops* and *Sicus*, Kempston Garden, Bournemouth West, vii. 15–17. See *Typical Flies*, Series 1, figs. 121, 123, 124. Sun-loving flies, affecting garden flowers, specially white ones.

Fig. 82. *Myopa polystigma* Rnd. ♂ 6 × 14 mm. Specimens found on apple blossom at Gt. Waldingfield, near Sudbury, Suffolk, v. 22. (Harwood.) A dark brown fly with brown legs and brownish wings throughout.

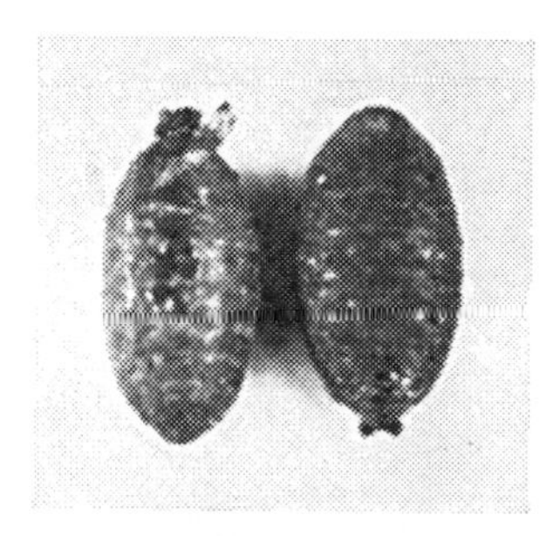

Fig. 83. Pupae of *Blepharidea vulgaris* Fln. 6 × 4 mm. ex *Vanessa urticae*. In certain hosts the parasites are smaller, perhaps caused by varying supplies of food. See *Typical Flies*, Series 2, fig. 68.

Fig. 84. *Tachina larvarum* L. ♀ 11 × 18½ mm. Pupa case (end missing) 10½ × 4½ mm., red and smooth. Bred Harwood ex *carpini*, Wellington College, Berks. A grey-black fly, tessellated slightly, striped thorax. Pupa red-brown. The Tachinidae are parasitic flies, the hosts of many are still to be studied; and having as known hosts Coleoptera, Hemiptera, Hymenoptera, etc.

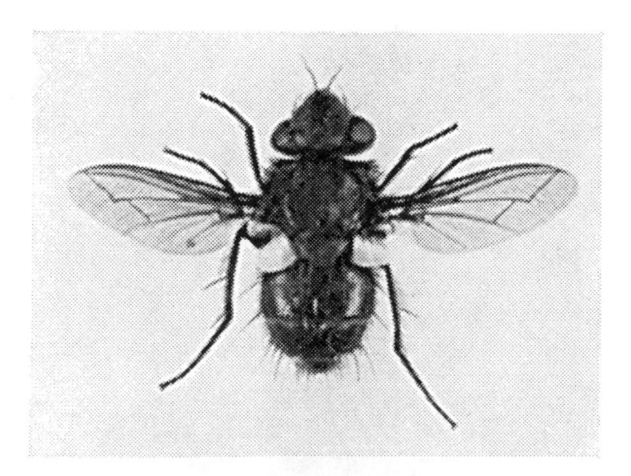

Fig. 85. *Gonia fasciata* Mg. ♀ 10 × 20 mm. A dark black abdomen with cross bars of white, and infumated wings.

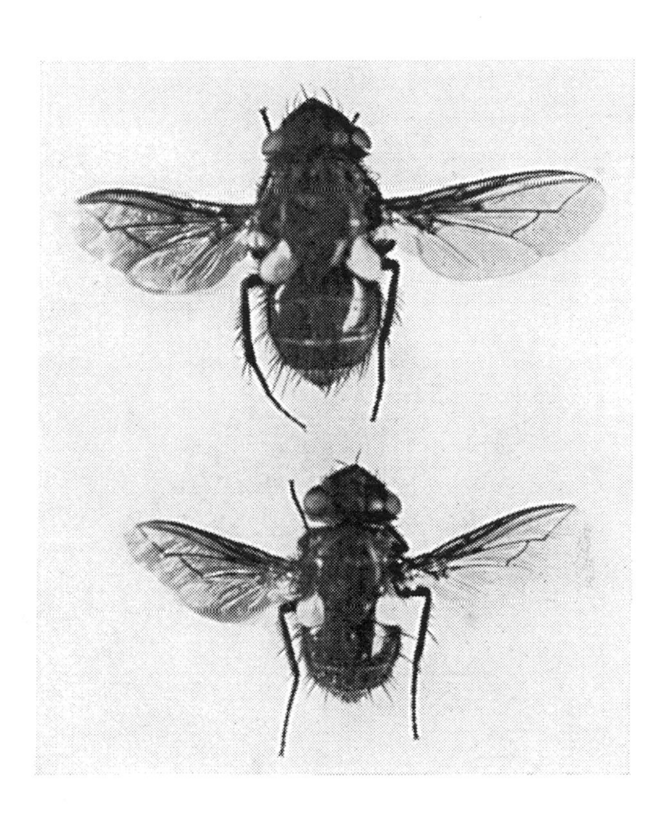

Fig. 86. *Gonia divisa* Mg. Two ♂♂ 11 × 21 mm., 10 × 17 mm. 7. v. 17. Haines, Dorsetshire. A red-brown fly with black longitudinal bar to abdomen and dense bristles.

Fig. 87. *Gonia ornata* Mg. ♂ 12 × 19 mm. ♀ $9\frac{1}{2} \times 16\frac{1}{2}$ mm. (♂ 21. v. 18; ♀ 26. iv. 12, heathland, Haines, Dorset.) A red-brown fly with longitudinal black stripe and two silver bands on abdomen. (Determined by E. E. Austen and Haines, Dorset.)

Fig. 88. *Micropalpus vulpinus* Fln. ♂ 10 × 15 mm. Grey thorax, red-brown abdomen flecked with silver. ♀ 9 × 17 mm. Grey thorax, abdomen red-brown, silver bands. On heathland, Wareham Bog, ix. 22. The rows of bristles on both sides of the frons are in a double row in the ♀, in a single row in the ♂. Frons wider in ♀.

Fig. 89. The heathland habitat of *Micropalpus*. Woolbarrow, Dorset. (Photograph, N. D. F. P.)

Fig. 90. *Trixa alpina* Mg. ♀ $12\frac{1}{2} \times 24$ mm. A grey fly with black markings, dense bristles and brown veined wings. 2. vii. 06. New Forest, Adams.

♂ ♀

Fig. 91. *Trixa oestroidea* Dsv. ♂ 11 × 20 mm. (Wiltshire, 1915. P. H.). ♀ 11 × 20 mm. (Wantage, 16. vii. 07). A red-brown abdomen somewhat densely bristled. A grey-black striped thorax and red-brown eyes.

Fig. 92. *Cynomyia mortuorum* L. ♂ 12×23 mm. A coast fly, though not entirely so, and not to be found every year. Sometimes not rare on *Aster tripolium*. This fly said to be common in Norway and attracted to fish when curing. (Theobald.) Dark thorax, vivid blue abdomen, black legs; frons yellowish.

Fig. 93. *Miltogramma punctatum* Mg. ♂ $5\frac{1}{2}$×9 mm. Pupa 7×3 mm. Parasitic on bee *Colletes danisana*, 8×15 mm. ♂ a grey fly, with three distinct black abdominal bands, dark thorax (♀ with fainter black splashes, grey thorax). Pupa red-brown. (Harwood.)

Fig. 94. *Colletes danisana*: the host of *Miltogramma*. These bees are gregarious and burrow in sand and soft banks or walls; this figure illustrates one of the colony from which the fly was obtained. Abdomen black with pale narrow bands of ivory colour and pubescence also. 8×15 mm.

Fig. 95. *Dexia rustica* F. ♂ 13×23 mm. Found near Peterborough and Royston, viii. 16. Rare. (Harwood.) A pale biscuit coloured fly with indistinct black stripe longitudinally on bristled abdomen, slightly infumated wings.

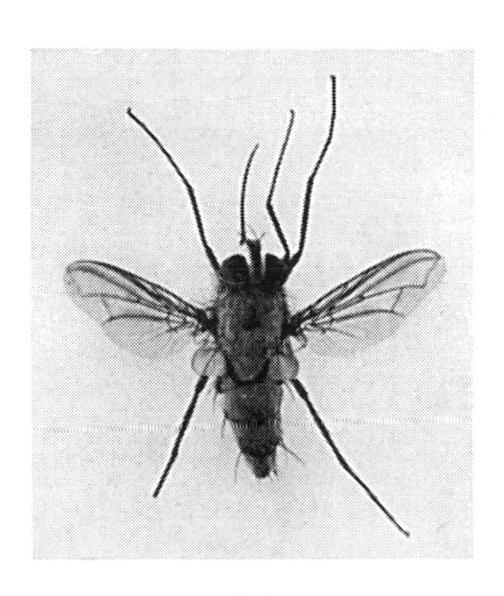

♂

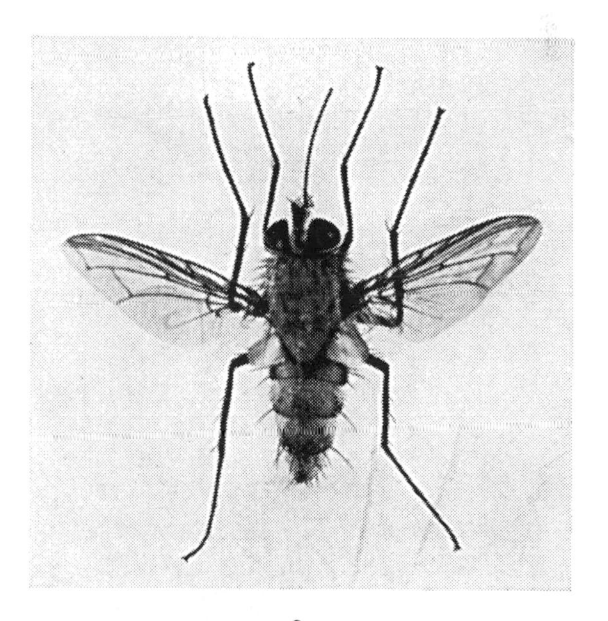

♀

Fig. 96. *Dexia vacua* Fln. ♂ 8×14 mm. ♀ 10×17 mm. On tree trunks near Tuddenham, W. Suffolk. A grey tint predominating, the abdomen yellowish with indistinct black stripe, black bristles, thorax grey and the frons silvered; a delicately coloured fly. (Harwood.)

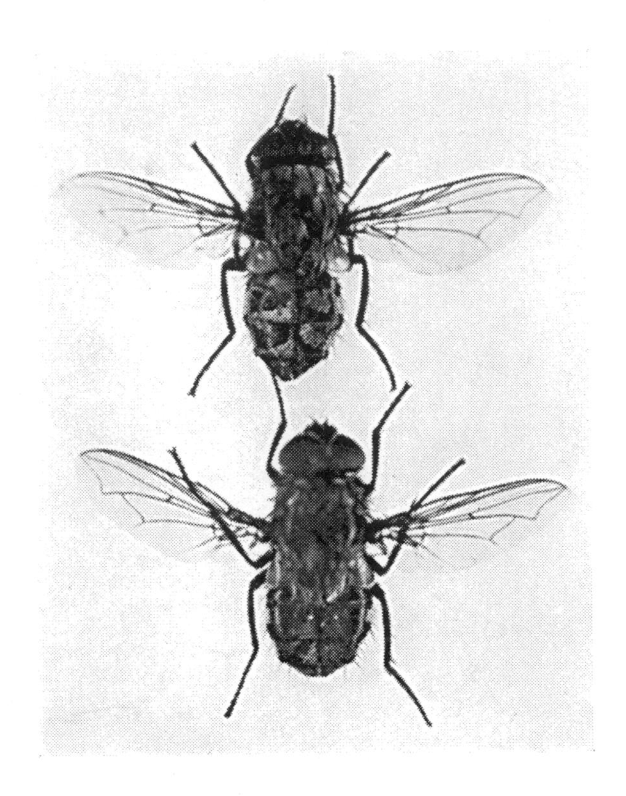

Fig. 97. *Pollenia rudis* F. ♂ 8×18 mm. abdomen tessellated, black legs. ♀ 9×18 mm. more decided black markings, bristled sides of abdomen. Has been known to hybernate in key holes and behind window sashes. "The cluster fly" from its habit of congregating in clusters in dwellings. (Harwood.)

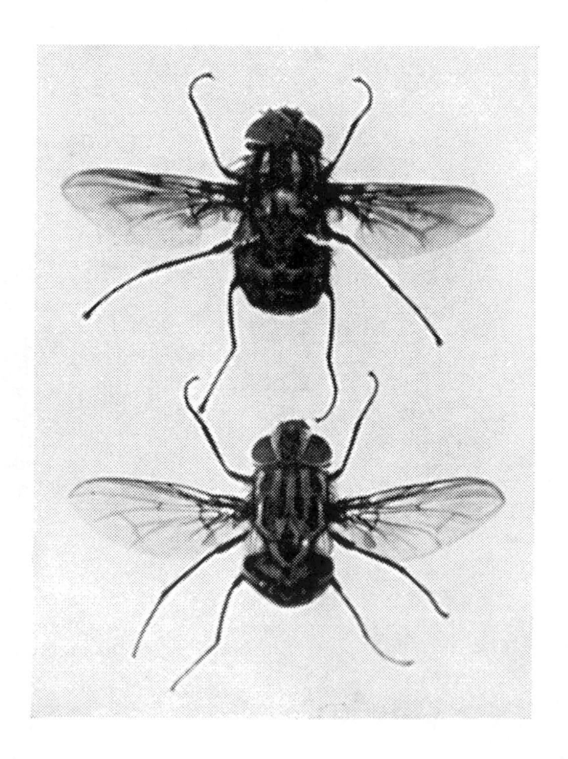

Fig. 98. *Graphomyia maculata* Scop. ♂ 8½ × 19 mm. ferruginous wings, yellow abdomen. ♀ 8 × 19 mm. clear wings, grey tessellated abdomen and striped thorax. (Grantchester, at Asters, prevalent in gardens, September. N. D. F. P.) The larvae live in cow and horse dung. (Theobald.)

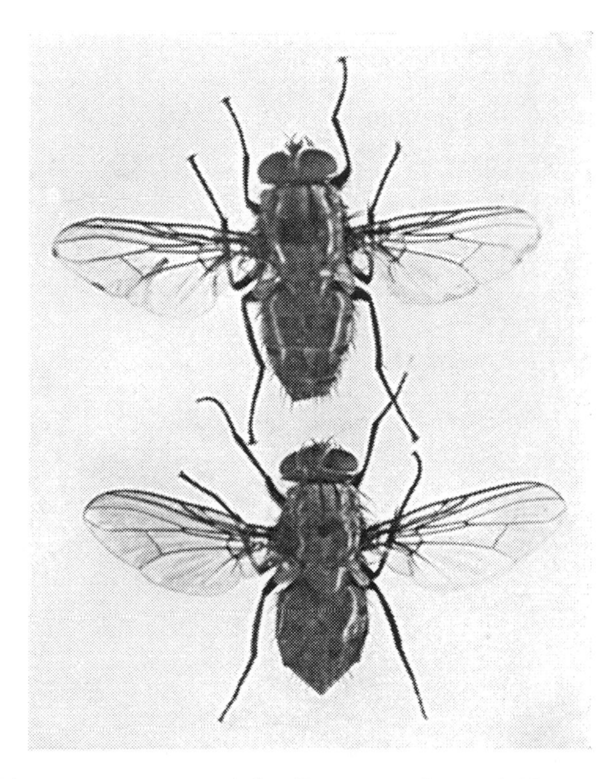

Fig. 99. *Cyrtoneura stabulans* Fln. ♂ 9 × 18 mm. ♀ 8½ × 17 mm. Dark grey with heavy bristles on sides of abdomen, brown legs. Found on windows from April. A common household fly.

Fig. 100. *Mesembrina meridiana* L. ♂ 12 × 27 mm. (reversed specimen showing undersurface). ♀ 12 × 25 mm. found in pasture lands and elsewhere sunning themselves on trees, wings luteous at base and foreborder. A black fly, comes freely to "sugar." Seen on banks of the Tay; on trees near Coupar Angus in July.

Fig. 101. *Calliphora erythrocephala* Mg. ♀ 10 × 22 mm. Pupa case 7 mm. The well-known "Blue-bottle Fly." One of the Blow Flies.

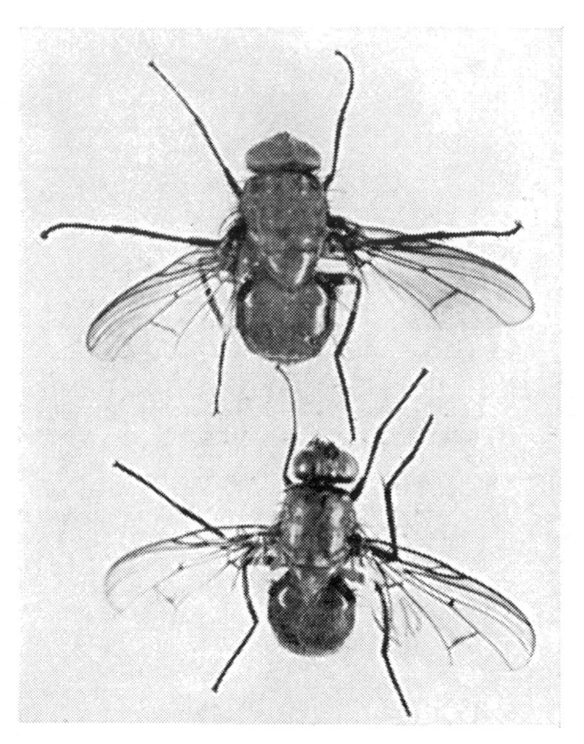

Fig. 102. *Hyetodesia scutellaris* Fln. ♂ 8½ × 17 mm. ♀ 8 × 17 mm. On windows of Kempston, Bournemouth, near ivy.

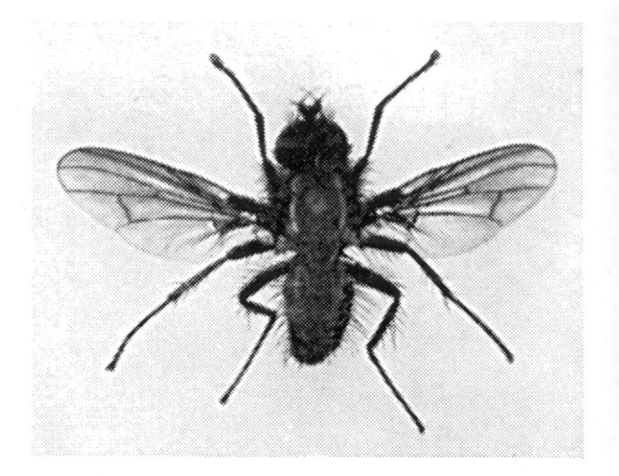

Fig. 103. *Drymia hamata* Fln. ♂ 7 × 14 mm. Stradbally. Brownish wings, sides of abdomen with fringe of bristles: thorax and legs with many bristles. A dark brown fly.

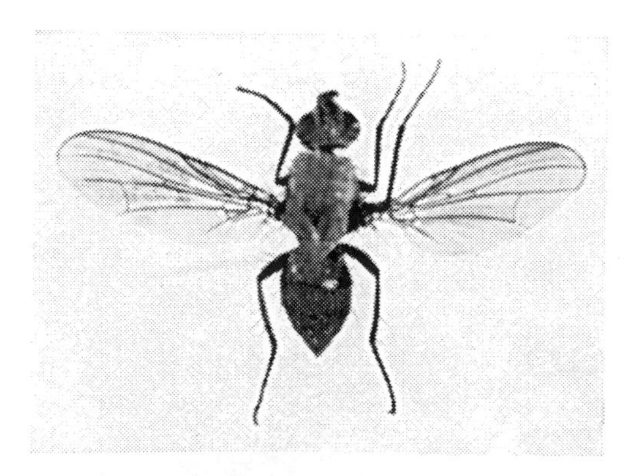

Fig. 104. *Leptohylemyia coarctata* Fall. (*Hylemyia coarctata* Fln.) ♀ 6 × 14 mm. The Wheat Bulb Fly, very destructive to wheat. Pale brown throughout, abdomen with bristles. (Theobald.)

Fig. 105. *Phorbia lactucae* Bouché. ♂ 5 × 12 mm. ♀ 4 × 10 mm. Pupa 5 mm. The well-known Lettuce Fly, destructive to seeds. ♂ dark brown, ♀ with grey abdomen; both with brown legs, brownish wings; pupa red-brown and elliptical in shape.

Fig. 106. *Phorbia cepetorum* Meade. Reversed specimen showing underside. ♀ 7 × 14 mm. Pupa case 7 mm. The Onion Fly. See *Typical Flies*, Series 1, fig. 144. A grey fly with distinctive bristles on thorax; bright brown pupa.

Fig. 107. (*Pegomyia betae* Curt.) ♂ 5 × 12 mm. ♀ 5 × 12 mm. Pupa 5 × 2 mm. From *Beta maritima*. Clacton-on-Sea. Harwood. A grey fly throughout; bright red-brown pupa. Often a very serious pest to mangolds and sugar beet in the field and to garden beetroot. The larvae mine and blister the leaves.

Fig. 108. *Pegomyia nigritarsis* Ztt. ♀ 5 × 10 mm. Pupa $5\frac{1}{2}$ × 2 mm. From Dock, of which it mines the leaves. (Harwood.) Abdomen amber colour, legs also, thorax grey. Pupa red-brown.

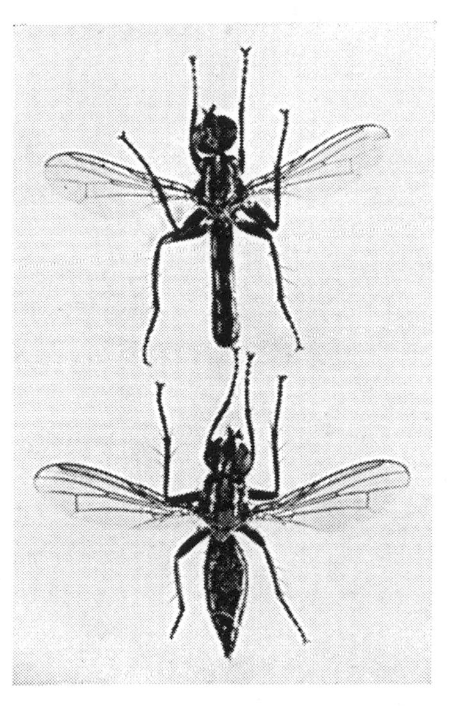

Fig. 109. *Cordylura pubera* F. ♂ 9 × 14 mm. ♀ 9 × 15 mm. A brown-grey fly with exceeding bristly legs and lightly veined brown wings, legs chiefly orange to pale brown.

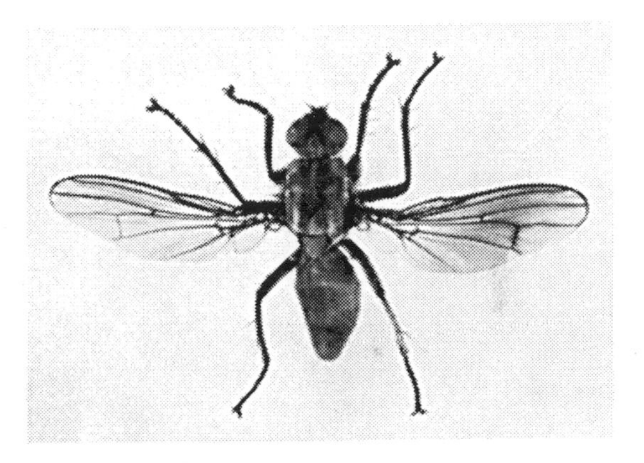

Fig. 110. *Scatophaga oceana.* ♀ 8 × 16 mm. (St Osyth Coast. Harwood.) (*Coratinostoma ostrorum* Hal.) Just above high water mark, just below as the tide goes down. A dark brown fly throughout with brownish wings.

Fig. 111. *Fucomyia frigida* Fln. ♂ 5½ × 11 mm. ♀ 6 × 14 mm. In thousands beneath seaweed at Millhook, Cornwall, July 1918. (Harwood.) A dark brown fly. The ♂ with many more bristles than the ♀ on legs and sides of abdomen (not visible in plate).

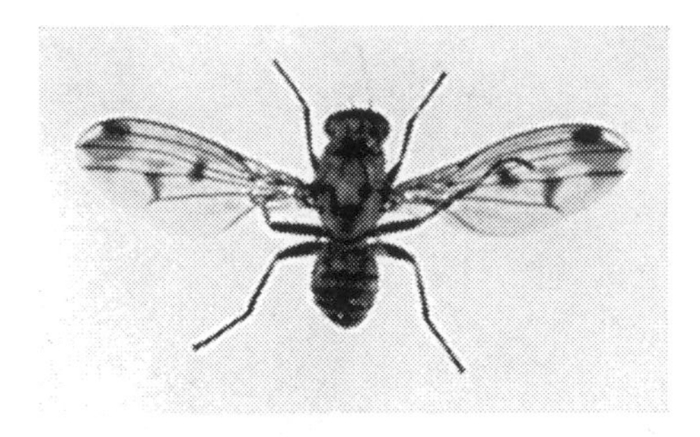

Fig. 112. *Helomyza rufa* Fln. ♀ 5 × 12 mm. New Forest. A brown fly with narrow dark brown bands on the abdomen.

Fig. 113. *Helomyza pallida* Fln. ♂ 6 × 15 mm. ♀ 5 × 12 mm. A fawn coloured fly of luteous abdomen bearing dark stripes; wings slightly infumated.

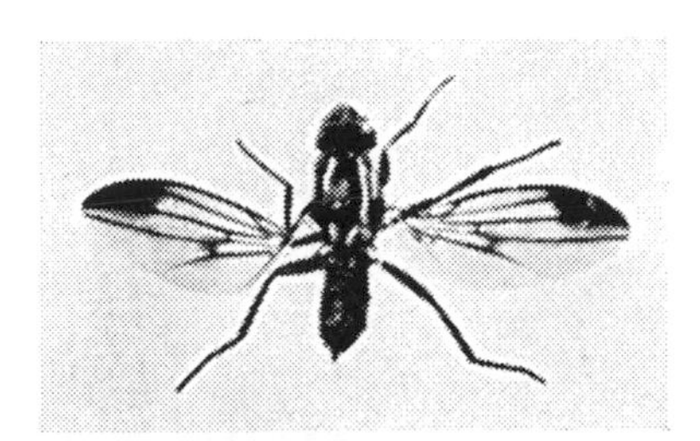

Fig. 114. *Heteroneura albimana* Mg. ♀ $4\frac{1}{2}\times9$ mm. Williston notes the larvae have the power of leaping (as those of *Piophila*) and are white, found in decaying wood. A black fly with dark brown blotches on apex of wings.

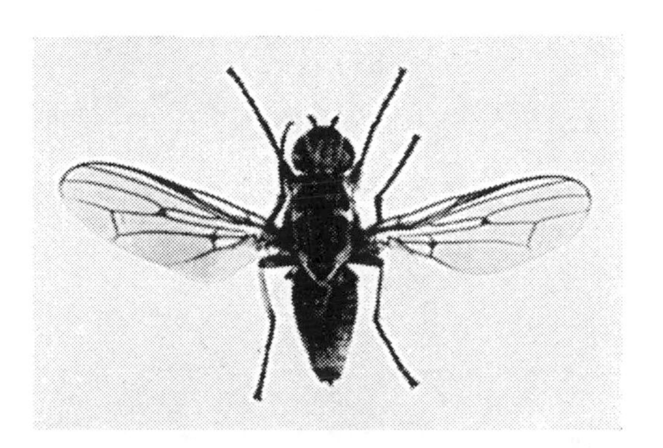

Fig. 115. *Actora aestuum* Mg. ♀ (?) 9×30 mm. A large shore fly. (Harwood.) Smooth and grey in colour with brown veined wings. See Meigen, vol. III, p. 403. ♀ from England. ♂ from Jutland, more stumpy than ♀ and with hairy forelegs.

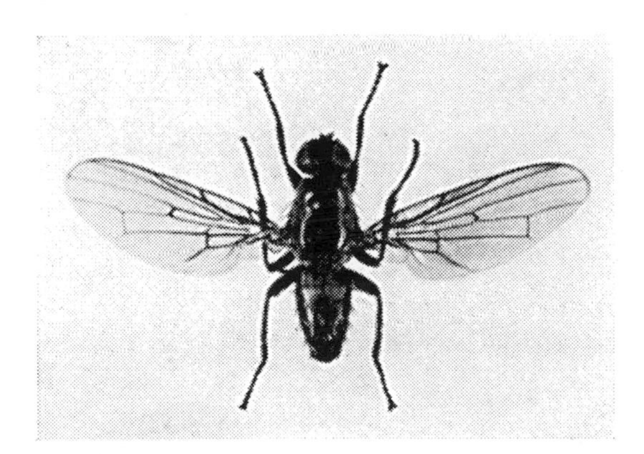

Fig. 116. *Dryomyza flaveola* F. ♀ 9×22 mm. An ochreous brown fly with distinctive black bristles on the thorax and the abdomen fringed with hairs. Shady woods.

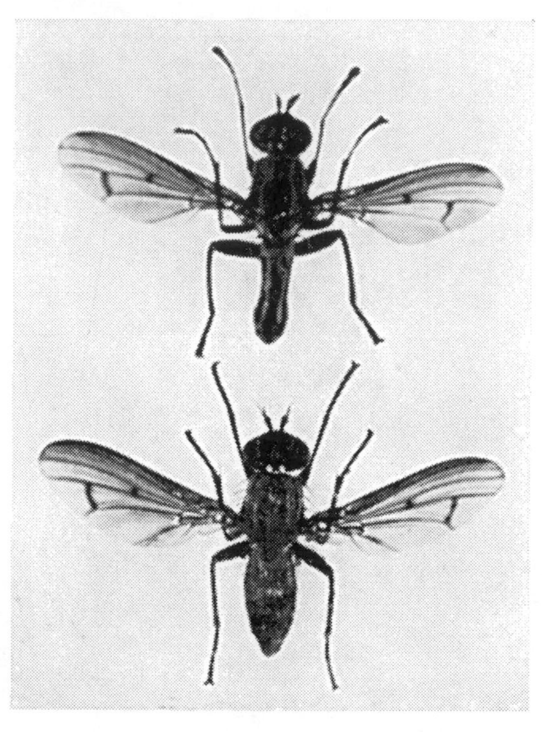

Fig. 117. *Tetanocera ferruginea* Fln. ♂ $8\frac{1}{2} \times 17$ mm.: light red-brown throughout. ♀ $8\frac{1}{2} \times 18$ mm. The ♂ has a few black bristles between the antennae.

Fig. 118. *Tetanocera coryleti* Scop. ♀ $9\frac{1}{2}$ mm. frons to extremity of wing. ♀ 8×17 mm. A pale brown fly with tracery of wings of same hue and bristles at extremity of abdomen.

Fig. 119. *Sepedon sphegeus* F. ♀ $5\frac{1}{2} \times 12\frac{1}{2}$ mm. A red-brown shining abdomen and legs, thorax being grey, infumated wings with spotted appearance, yellow brown.

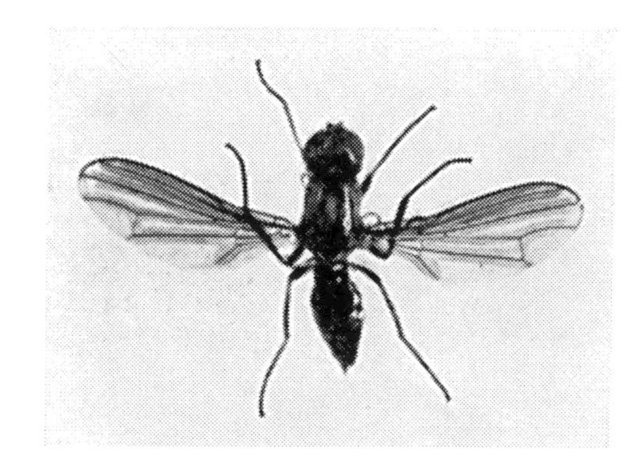

Fig. 120. *Psila rosae* F. ♀ $5 \times 10\frac{1}{2}$ mm. A smooth dark brown shiny fly with yellow head, devoid of hairs or bristles. "The Carrot Fly," producing the rust in carrots. It also attacks Parsnips. (Theobald.)

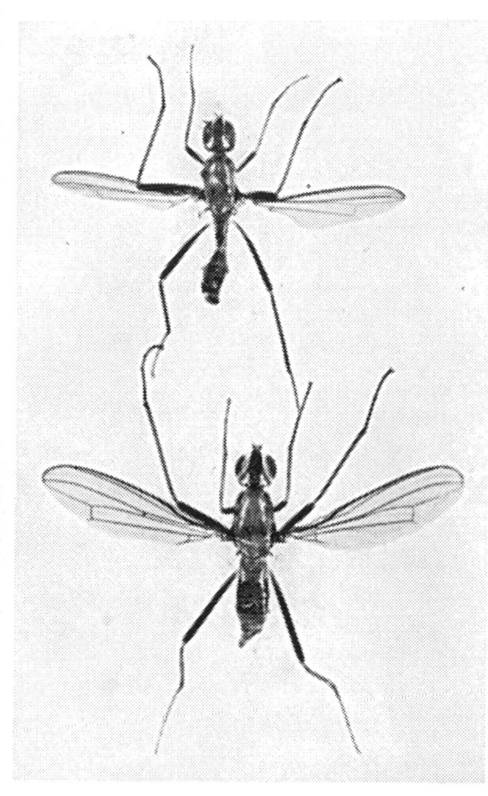

Fig. 121. *Calobata cothurnata* Pz. ♂ $6\frac{1}{2} \times 13$ mm. ♀ $7\frac{1}{2} \times 14\frac{1}{2}$ mm. A smooth dark brown fly with pale yellow-brown legs, and curiously spherical head.

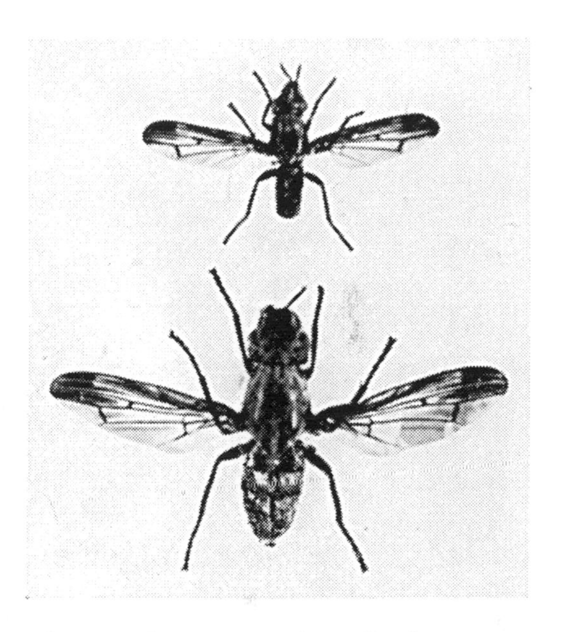

Fig. 122. *Dorycera graminum* F. ♂ 5×11 mm. (the ♂ varies greatly in size). ♀ 9×16 mm. Not generally common but local; beaten from Wych Elm. A light brown fly, thorax striped dark brown in ♀, abdomen banded in dark brown hue.

Fig. 123. *Pteropaectria afflicta* Mg. 5×10 mm. Wicken Fen, N. D. F. P. (Not common, Adams.) A very dark brown fly, legs same, and brown splashed wings.

Fig. 124. *Ulidia erythropthalma* Mg. 4×8 mm. A black shining smooth fly, brownish wings.

Fig. 125. *Acidia heraclei* L. 5 × 12 mm. Pupa 5 × 2 mm. Mines leaves of celery plants, often a severe pest. Dark brown markings of wings, pupae pale buff, often coloured by earth adhering.

Fig. 126. Leaf of celery, portion of which shows blister containing mining larvae of *A. heraclei*.

Fig. 127. *Spilographa alternata* Fln. ♂ 5 × 12 mm. ♀ 7 × 14 mm. Larvae breed in rose hips which only show attack by a dark discoloration; pupates in the ground. Many seen near Halstead, Essex. (Harwood.) A pale yellow-brown fly with wing bars of dark brown.

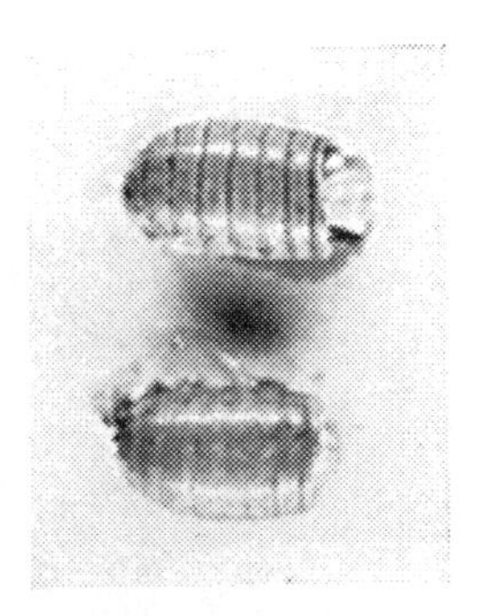

Fig. 128. Pupae of *Spilographa alternata* Fln. Apex gone, 4 mm. pale cream colour.

Fig. 129. *Urophora macrura* Lw. ♀ 7½ × 12 mm. N. D. F. P., Devil's Dyke, Swaffham. On *Onopordum illyricum* L., a composite plant. "Réceptacle épaissi et déformé." A dark brown fly with wing markings of same colour.

Fig. 130. *Urophora quadrifasciata* Mg. ♀ 4 × 6½ mm. Black abdomen and thorax, wings dark brown stripes on white ground. On *Centaurea Jacea* L. (Brown Radiant Knapweed), *C. nigra* L. and *C. paniculata*, etc. "Réceptacle floral, transformé en une cécidie cylindrique, subconique ou ovoïde, très dure, multiloculaire." (Second and third bands not united at hind margin of wing.)

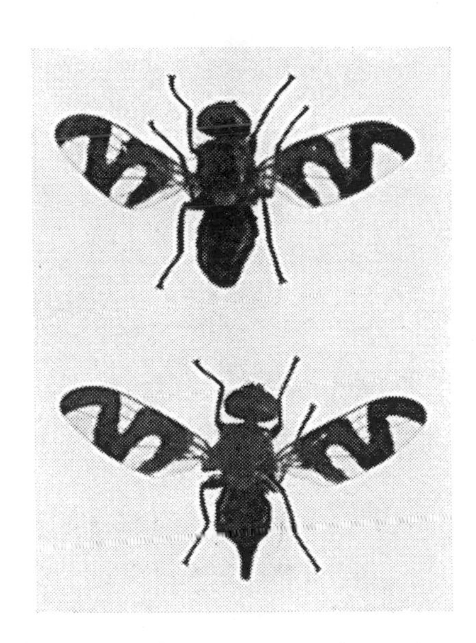

Fig. 131. *Urophora cardui* L. ♂ 5 × 10½ mm. ♀ 7 × 11 mm. From thistle galls (bred Harwood). Forms galls on *Cirsium oleraceum* and *C. arvense*. Wings with dark brown striping on white ground, black bodies. Second and third bands united at hind margin of wing.

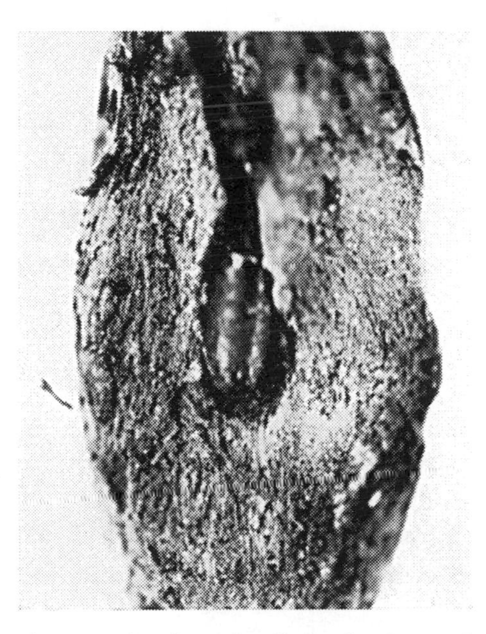

Fig. 132. Section of gall showing pupa of *Urophora cardui* in situ, exit shown, pupa 4½ mm. in gall.

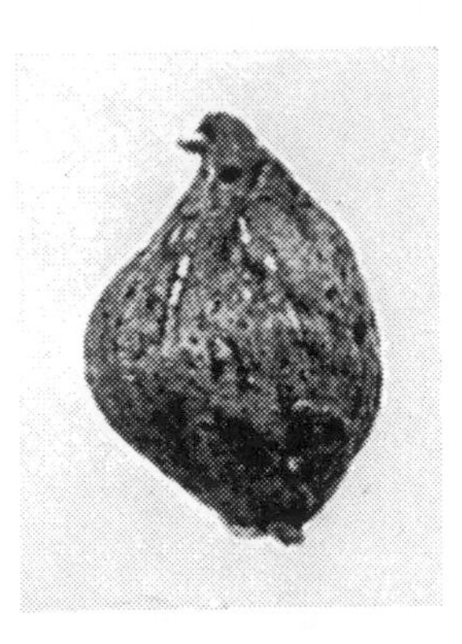

Fig. 133. Entire gall, actual size about 27 × 19 mm.; from thistle, dark grey in hue. "Réceptacle floral hypertrophié, transformé en une cécidie pluriloculaire de consistance normale. Renflement arrondi, ou fusciforme, dur, multiloculaire, dépassant le plus souvent la grosseur d'une noix et pouvant atteindre la dimension du poing." From Honard, *Les Zoocécidies.*

Fig. 134. *Icterica Westermanni* Mg. ♀ 5½ × 12½ mm. Sway and Isle of Wight. (Adams.) A red-brown fly with white markings.

Fig. 135. *Ensina sonchi* L. ♀ 3 × 8 mm. In sow-thistles. (Harwood.) A grey thorax, abdomen darker, slightly narrow bands of yellow.

Fig. 136. *Tephritis miliaria* Schrk. ♂ 6 × 13 mm. ♀ 10 × 14 mm. 8. vii. 09. New Forest, Adams. A yellow-brown fly with darker brown blotches on brown veined wings.

Fig. 137. *Tephritis absinthii* F. ♂. bred by Harwood from *Artemisia maritima.* A dull brown fly with blotches of lighter hue.

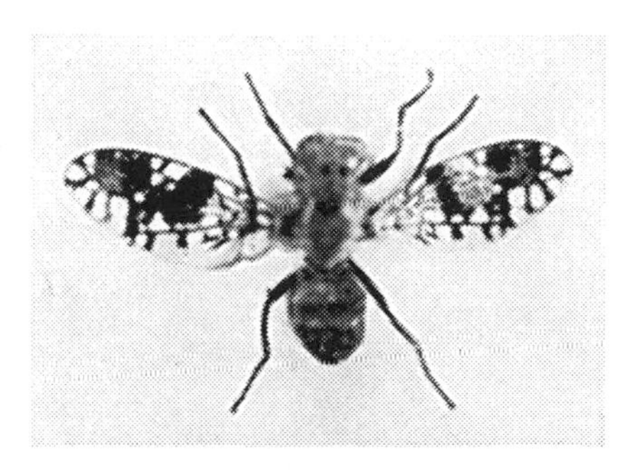

Fig. 138. *Tephritis vespertina* Lw. ♂ $3\frac{1}{2} \times 8$ mm. A brown fly with wings reticulated, legs dark brown.

Fig. 139. *Tephritis parvula* Lw. ♀ 5×10.

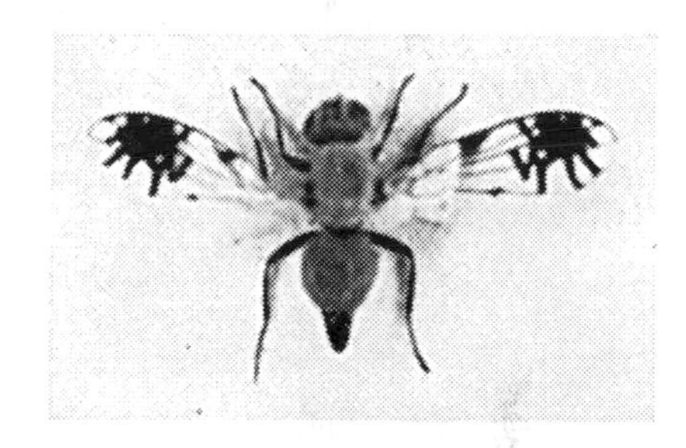

Fig. 140. *Urellia amoena* Frfld. ♀ $3\frac{1}{2} \times 7$ mm. "A star-shaped black picture at apex of wings," Williston (who attributes it to Desvoidy).

Fig. 141. *Lonchaea vaginalis* Fln. ♂ $4\frac{1}{2} \times 10\frac{1}{2}$ mm. ♀ $4 \times 10\frac{1}{2}$ mm. Shiny black fly (smooth), wings brown veined, irridescent.

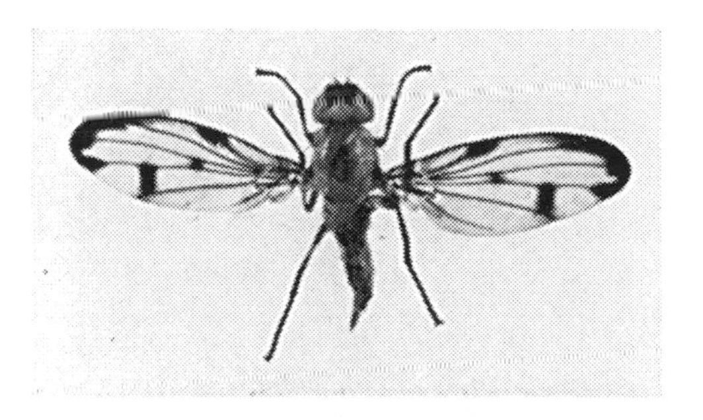

Fig 142. *Palloptera arcuata* Fln. ♀ 5×11 mm. A pale buff fly, black bristles on head and thorax, pale legs, darker wing marks, and abdomen.

Fig. 143. *Sapromyza inusta* Mg. ♂ 5 × 11. ♀ 5 × 12 mm. N. D. F. P., Grantchester and Kempston, Bournemouth, also at Colchester. (Harwood.) Buff thorax and legs, darker abdomen, wings spotted same shade, black bristles on thorax.

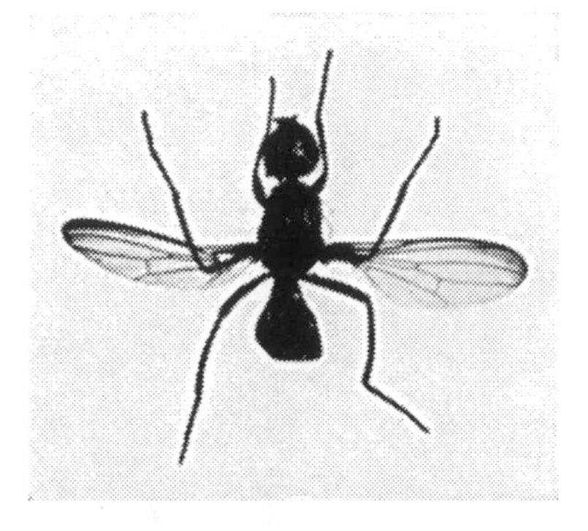

Fig. 144. *Nemopoda cylindrica* F. ♀ $4\frac{1}{2} \times 8\frac{1}{2}$. "Active flies, quick in flight," Williston. Head rounded and bristly; a dark brown fly, brown legs, brown veined wings.

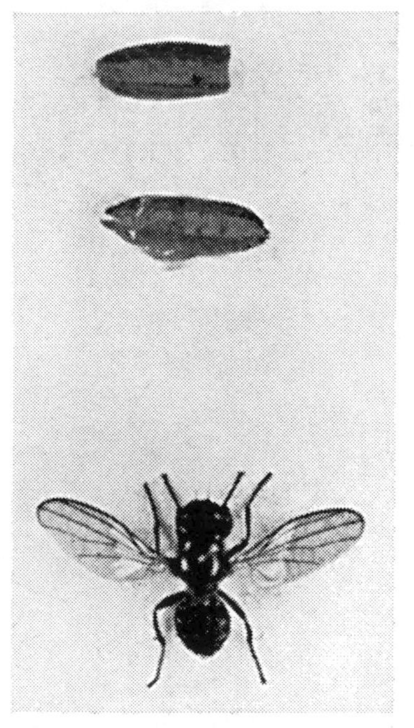

Fig. 145. *Piophila vulgaris.* ♂ 3 × 7 mm. Pupa 3 mm. Puparia found in and under hard fungi, Clacton-on-Sea, the fungus being thrown up by the tides. iv. 1912. (Harwood.) A small black fly, pupae bright red-brown.

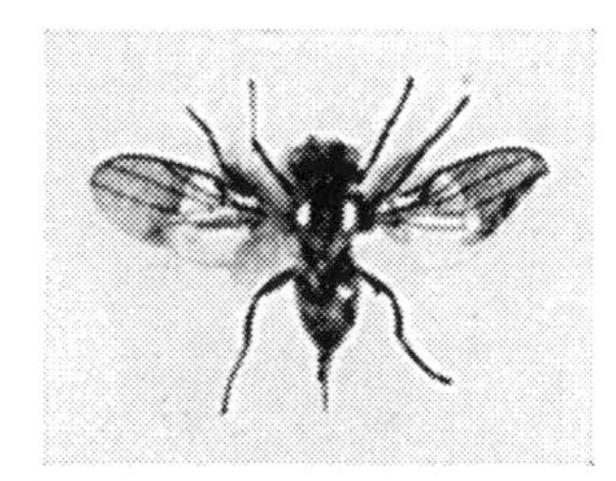

Fig. 146. *Oscinella* (*Oscinis*) *frit* L. The Frit Fly. ♀ 2 × 4 mm. to end of ovipositor. Black fly with skipping habits. The larvae are very destructive to Oats all over Europe. There are two broods, one attacks the young plants, the second the grain. Young plants may be killed and the grain is shrivelled and spoilt.

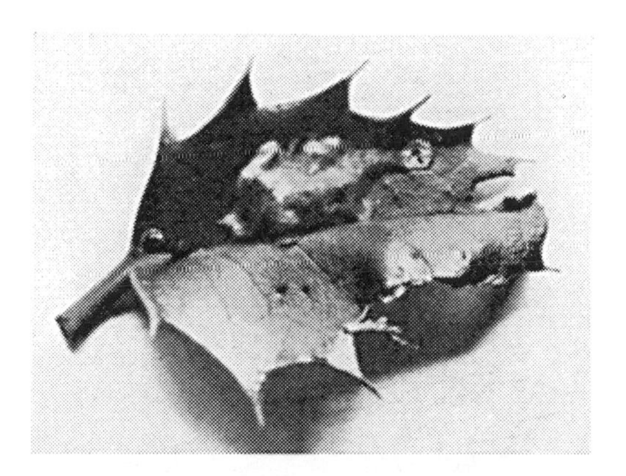

Fig. 147. Leaf of garden holly mined in blotches by *Chromatomyia ilicis* Curt. Very harmful to hollies, the blotch made being 20 mm. in length: the larva pupates therein, emerging in June. See *Typical Flies*, Series 2, fig. 109. (Kempston Holly Hedge, Bournemouth.)

Fig. 148. *Stenopteryx hirundinis* L. 4 × 9mm. Puparia 2 × 3mm., spherical and dark brown, in nests of Martins. Hauxton Mill, N. D. F. P. ix. 16. Bred out at Bournemouth, v. 17. E. K. P. The puparia seem to lie dormant in the nests from autumn until the return of the Martins in spring. A pale brown fly. See *Typical Flies*, Series 2, figs. 115, 116.

APPENDIX

(*The Specimens in the Appendix are not arranged consecutively*)

Fig. 149. *Xiphura atrata* L. ♀ 31 × 22 mm. Upper half of abdomen luteous orange (*nigricornis* has a brown band on orange portion of abdomen). Glemmorew.

Fig. 150. *Zodion cinereum* F. 6 × 11 mm. Greyish abdomen, speckled black. The proboscis straight, not elbowed as in other members of the family, or rather elbowed at the base: darkish wings, dark legs. (Harwood.)

Fig. 151. *Oncomyia atra* F. 6 × 10 mm. (Harwood.) A black fly, proboscis elbowed in middle.

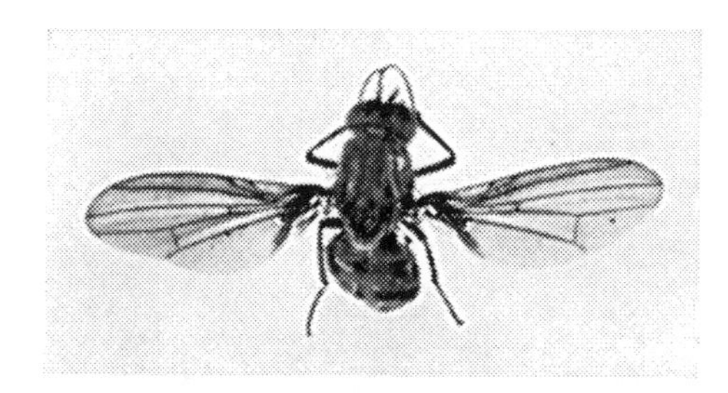

Fig. 152. *Sapromyza longipennis* F. ♂ 4 × 11½ mm. A dark fly with dark brown colouring, decided buff wings and few veins.

Fig. 153. *Thyridanthrax (Anthrax) fenestratus* Fln. (The Moorland Fly.) ♂ 9×22 mm. Second and third segments of abdomen chestnut colour. ♀ no chestnut colouring and head wider. *Anthrax fenestratus* Fall. (Verrall). Dr Haines, who has a series taken on the heath country beyond Wareham, Dorset, says that they are nearly all ♂ : ♀ is rarely taken. On the sandy road near Sherford Bridge (Wareham and Morden Heath) I myself saw twelve, of which I took six or seven, all apparently ♂. In movement it is a very snipe among flies, here to-day and gone to-morrow. It is often much rubbed in taking and boxing, unless immediately killed or "slept." The late Mr Adams found a colony on Umbelliferae. Generally they bask on hot sandy paths and roads near bogs (Morden Park Lake upper bridge, Matley Bog, New Forest in a sandpit). Adams took them in the New Forest. Mr Lamb says that it is impossible to show in a photograph the distinction between the sexes. Adams says that he took also *Villa circumdata (Anthrax circumdatus)* on bloom of *Heracleum Sphondylium* at Matley, vii. 99, as well as on *Angelica sylvestris*, and close to his house, Fern Cottage (White Meadow), when he lived in the New Forest.

Fig. 154. Wareham Bog in winter, the summer haunt of rare flies: foot of Great Ovens. (Photograph, J. Pearce.)

Fig. 155. *Lipara lucens* Mg. ♂ 5×12 mm. ♀ 6×12 mm. Chippenham Fen. Bred vi. 27. (B. Harwood.) "*Lipara* is thought to oviposit in a young green reed; quite possibly on the outside, or at the top and the young larva to eat in; the hole, as is the case with those made in young apples by *Hoplocampa*, would be quite invisible": the *Lipara* drops and feigns death on approach of the net (Harwood); this is also I believe stated by Curtis, and makes it almost impossible to effect a capture. So the fly must be bred.

Fig. 156. Galled reed of *Lipara lucens* (Harwood, Chippenham Fen). Length of reed from where holed by some small insect (perhaps *Prosopis*) to top of reed 38 mm., circumference 23 mm. The aborted knob-like inflorescence at top of gall indicates that the larva on hatching burrows downward.

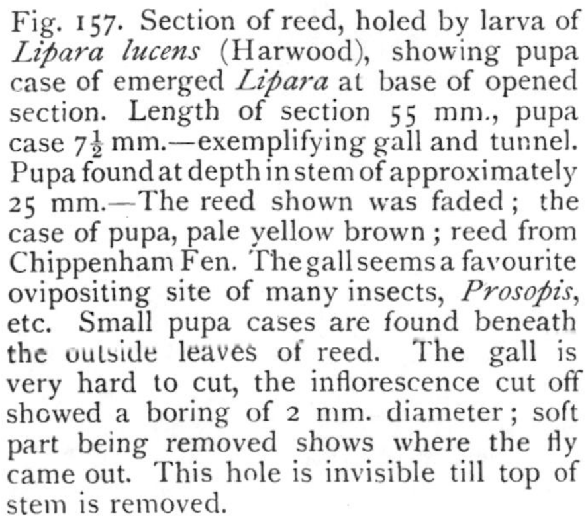

Fig. 157. Section of reed, holed by larva of *Lipara lucens* (Harwood), showing pupa case of emerged *Lipara* at base of opened section. Length of section 55 mm., pupa case $7\frac{1}{2}$ mm.—exemplifying gall and tunnel. Pupa found at depth in stem of approximately 25 mm.—The reed shown was faded; the case of pupa, pale yellow brown; reed from Chippenham Fen. The gall seems a favourite ovipositing site of many insects, *Prosopis*, etc. Small pupa cases are found beneath the outside leaves of reed. The gall is very hard to cut, the inflorescence cut off showed a boring of 2 mm. diameter; soft part being removed shows where the fly came out. This hole is invisible till top of stem is removed.

Fig. 158. *Digonochaeta setipennis* Fln. $9 \times 4\frac{1}{2}$ mm. (Harwood.) Lays its eggs on earwigs (usually one, occasionally two eggs) and choosing in all probability the young insects; when full fed the larvae emerge from the earwig. May pupate anywhere, in hollow stems, or behind loose bark; found by Harwood even in muslin "sleeves" of caterpillars. He has not found it in the earth; being so small it may not have been seen.

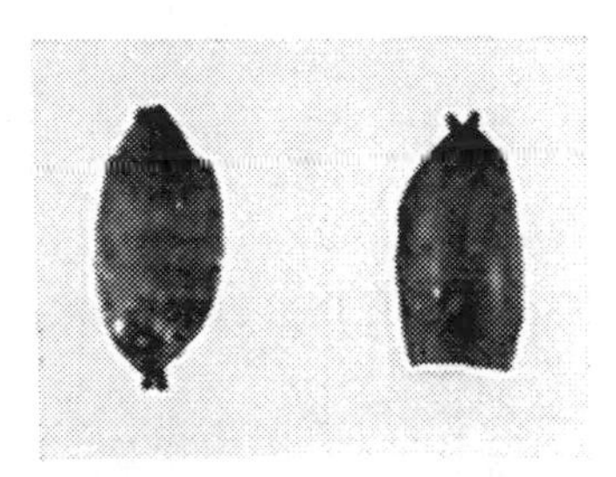

Fig. 159. *Digonochaeta setipennis* Fln. Pupae cases 4 mm., $4\frac{1}{2}$ mm. (Harwood.) The puparia seem subject to parasites; one was full of Chalcid maggots, from another a small *Ichneumon* was bred.

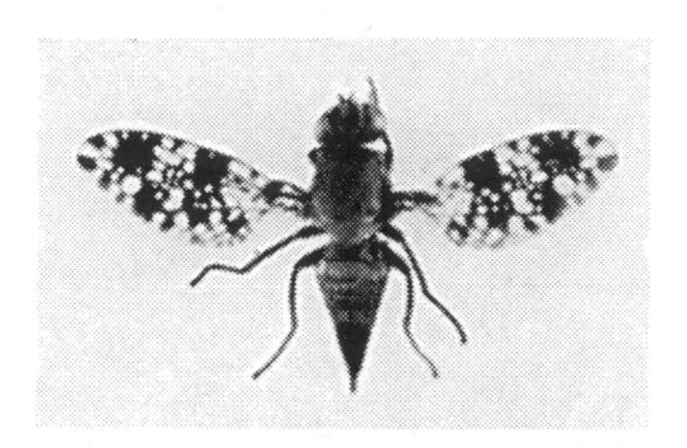

Fig. 160. *Tephritis (Euaresta) conjuncta* Lw. 9. v. 02. F. C. Adams. ♀ 4×9 mm. By permission of British Museum. Abdomen and base of thorax lightly fringed with greyish hairs giving a soft appearance to the general contour.

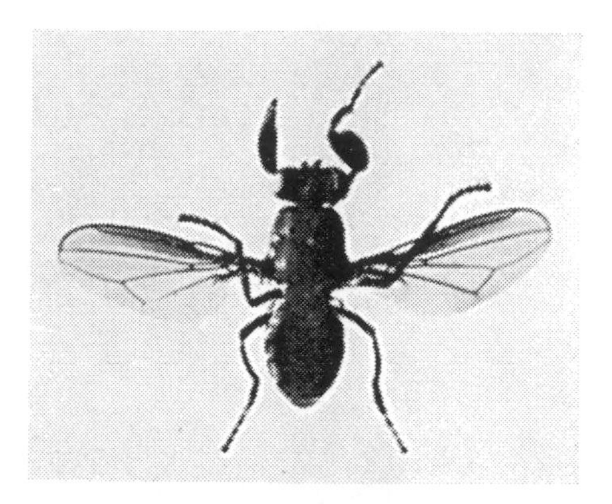

Fig. 161. *Ochthera mantis* Deg. Dorchester, Harwood. viii. 27. 5×10 mm. "Les cuisses antérieures dilatées en larges mains. Les Ochthères se trouvent sur les plantes aquatiques aux mois d'Août et de Septembre," Macquart. Described also as sipping drops of dew off the fore legs, and possibly of cannibalistic tendencies (?) by another description. A curious and probably unique (British) fly.

Fig. 162. *Tabanus sudeticus* Zlr. ♀ $21\frac{1}{2}$×41 mm. Nethy Bridge, vii. 25. (P. Harwood.) Thorax dark, darker abdominal bands and general appearance than *T. bovinus* and the white triangles more heart shaped, wings darker. Verrall points out upper facets of eyes larger in *sudeticus*. The largest British fly; Verrall and Austen also note it at Nethy Bridge. Adams does not include it in New Forest captures.

INDEX

The references are to the Figures